"THE DOUBLE PLANET"

"THE DOUBLE PLANET"

PLUS
"THE LIVES OF THE ELECTRONS"
PLUS
"CULTURE AND BELIEF"

Les Crane

iUniverse, Inc.
New York Lincoln Shanghai

"THE DOUBLE PLANET"
PLUS
"THE LIVES OF THE ELECTRONS"
PLUS
"CULTURE AND BELIEF"

iUniverse books may be ordered through booksellers or by contacting:

iUniverse
2021 Pine Lake Road, Suite 100
Lincoln, NE 68512
www.iuniverse.com
1-800-Authors (1-800-288-4677)

ISBN: 978-0-595-44629-2 (pbk)
ISBN: 978-0-595-88954-9 (ebk)

Printed in the United States of America

CONTENTS

INTRODUCTION

What follows are two short essays and one longer work that, in essence, concern three aspects of the same subject. That subject is the tantalizing question that has perplexed mankind from the beginning of human existence. That problem is the Great Mystery; the Mystery of life on this planet; the Mystery of how we might best organize ourselves in harmony with the Mystery; and the Mystery of the vast Universe we see in the starry skies above us.

This seeking after final answers has always been with us. Essentially, all the ancient religions were attempting to penetrate that Mystery; to reveal it to us in human terms. It is the essential element in all religions; both past and present.

Egyptians believed life was only a way-station between the unknown realm they had lived in before and the one they would inhabit in the future. Buddhists sought Nirvana as a way of melding with the Mystery. Jews followed strict codes of behavior as a way of justifying themselves before their unseen God. Savages in the jungles and on the plains, throughout the world, had the same need; they, too, felt the attraction of the Mystery. They could not penetrate it, nor can we, but they knew it was there and, in their ceremonies, placated it.

Modern society and science now believe they are on the brink of solving most of the unanswered questions and thus finally mastering the secret of the Mystery. But is this a vain hope? It is only during the last 600 years that we have even begun to free ourselves from the chains of tyranny and superstition; and been able to live in somewhat rational and free societies. In that short span of time, we have learned much; but we have more to learn.

Over those 600 years, science and technology has progressed from a primitive level; from a world lighted only by candles at night, into the development of chemistry, mathematics, medicine and electricity; and into a world lighted at night by electric power from atomic power plants. Along the way, during these titanic advances, we have attempted to codify our knowledge by creating scientific and political "Laws;" "Laws" that should be understood as only guideposts to our future advances in knowledge. These "Laws" are useful but not immutable and are subject to amendment as time goes on. Yet, the Mystery remains. It is a Force of Attraction that draws us forward to make up, ever new, theories about Creation and drive our space programs onward.

It is, in an attempt to analyze where we stand today, both scientifically and philosophically, that the following works are offered. If we fail to clearly recognize our precarious situation, this will cause us to drift back into that slough of igno-rance from which we have, only recently, freed ourselves. If our advances in sci-ence and philosophy are allowed to lapse and be forgotten, our civilization will self-destruct. This is not an idle threat. Great civilizations before us have tri-umphed and then collapsed into ruins. The world is littered with the broken monuments of their former greatness.

The following is a defense of the logic, reason and wisdom we have gained—so far; because the forces of unreason and tyranny are not idle. They are on the attack; the Barbarians are at the gates.

"THE DOUBLE PLANET"

Earth's Heat Engine

Les Crane

© 2007

Today, some believe in the disturbing possibility that the Earth may be entering into a "global warming" period. This concern has replaced their previous fears that Earth was beginning a "global cooling" trend. If either is true, are the activities of mankind seriously affecting Earth's temperature? And, if so, what can be done about it?

Before accepting either conclusion we should, perhaps first look at underlying factors that have regulated Earth's temperature in the past: those factors that made life upon Earth even possible. To do this, we must first ask what are the planet's underlying heating and cooling mechanisms.

Considered realistically, we and all other living things are inhabitants of a water-covered rocky sphere, floating utterly alone, in the limitless vacuum of interstellar space. Yet, despite this forbidding environment, Earth teems with life. The fact that life forms exist in so sterile an environment defies rational explanation. To the logical mind it would seem the Earth should be no more that a frozen rock; orbiting endlessly about a minor star; with its physical conditions being no different than that of the other, uninhabitable, inner planets, Mercury, Venus or Mars. But because, on Earth, organic life does exist; in truth, we cannot be alone—for there must God in some form. If one does not agree with religious conceptions, call it "The Force," or the "Little Green Men"—take your pick. But, the reality of life on Earth, forces us to accept a thing impossible—the dictionary definition of a miracle. Organic "Life," including yours and mine, truly, *is* a miracle; created by some, yet undiscovered, external power. Before this inexplicable fact, we must ever stand in wonder.

While the ultimate secrets of creation may never be revealed, we are still privileged to inquire into the mechanisms by which this miracle was accomplished and to consider a further question: "how does life on Earth continue to be preserved?" What are the underlying factors that maintain Earth's hospitable and temperate environment?

One generally unrecognized factor, controlling our climate, is that at some time after the Earth was set spinning upon its axis, another body appeared, or was somehow brought into being, in the form of a nearby second planet, the Moon. The Moon's gravitational attraction is fundamental to the existence of life on Earth because it is the Earth-Moon gravitational relationship that generates the latent heat energy that maintains Earth's temperature in its present state of equilibrium.

Contrary to common belief, the Moon is not really an Earth satellite. Although it is smaller than the Earth, the Moon functions, in truth, as if it were a second planet; acting in a symbiotic relationship with Earth; It is Earth's partner in a double-planetary system. That partnership might appear to be insignificant until we explore and consider its resulting physical effects. These include the fact that the Moon's gravitational attraction is the direct cause of oceanic tides. It is, also, the cause of earthquakes and provides the only force that could be creating the continental drift. More importantly, the Moon creates the twice daily Earth tides that work continually upon the body of the planet; an action that maintains Earth's temperature to within that essentially narrow range 'that permits the existence of organic life; the range between the freezing and boiling points of water; above or below which, biological life would cease to exist.

The Moon's orbit about the Earth is nearly circular and lies almost exactly upon the plane of the ecliptic, (the plane of the Earth's and all the other planet's paths about the Sun). This is puzzling because every theory so far advanced to account for the existence of the Moon in this orbit, has failed to provide any reasonable explanation as to why this should be. One theory asserts that the Moon may have been thrown off by the Earth during its earliest stage of formation. However, this theory has since had to be dismissed because, mechanically, it is impossible. Were Earth, to have ever rotated at the speed necessary to throw off the Moon, the rapidity of that rotation, with its accompanying centrifugal force, would have flung the Moon into outer space. Further, the Earth, at that speed of rotation, with its equatorial surface moving at over 17,000 miles per hour, would, itself, have broken into pieces or have never begun to form. A further difficulty is that the Moon's density has been found to be only one-half that of Earth. This argues against these two bodies ever having had a common physical origin.

The "Capture Theory" is another possibility. It attempts to explain the Moon's orbit by assuming it was, originally, a vagrant planetoid that, somehow, drifted into Earth's gravitational field and was, thereafter, permanently trapped as a satellite. But, had the Moon been captured in this manner, in all likelihood, it would have assumed an elliptical orbit—and, almost certainly, that orbit would have been steeply inclined to the plane of the ecliptic, as in the case of comets.

Since neither these theories seems acceptable, astronomers tend to avoid the question because their inquiries force them toward an answer they do not wish to consider; that the Moon may have been *placed* into its present orbit, either by divine fiat, or by some other intelligent agency!

TIDAL EFFECTS

The most easily recognized physical effect of the Moon is its control over the twice-daily tides in Earth's oceans. These oceanic tides are caused primarily by the gravitational attraction of the Moon although they are supplemented to a degree by the added attraction of the Sun. Depending upon their relative positions, the Sun either adds to, or subtracts from, the Moon's gravitational force by a factor of about 20%. At the phase of the new Moon, the Sun adds to the Moon's attraction while, at the full Moon, because, now, the two bodies are on opposite sides of the Earth, the Sun subtracts from the Moon's gravitational attraction upon the Earth.

While we know that the oceanic tides are caused by the Moon, the question next arises: "If the tides are primarily due to the attraction of a single body, the Moon, why are there *two* tides per day at any given point on the ocean's shores?" Would it not seem reasonable that a single object, such as the Moon, would create but one tide per day; drawing up the surface of the ocean's waters on the side directly facing itself as is shown in (Figure1)? However, what is actually observed is that there are *two* daily tides per day at any given location. These succeed each other at intervals of 12 hours and 20 minutes. The twenty minutes gained in each, twelve-hour period is because the Moon, itself, is traveling eastward in its orbit; in the same direction as Earth's rotation. Thus, to us, the Moon appears to rise, pass overhead, and set, forty minutes later each day. (Twenty minutes later in each twelve-hour period). Yet, the fundamental question remains. Why *two* tides?

The Encyclopedia Britannica offers the following explanation:

> "The separate attractions of the Moon, at the Earth's center, and at a point on the Earth's surface are each inversely proportional to the cube of the Moon's distance." [1]

This author argues that the gravitational force exerted by the Moon is much stronger at the near side surface of the Earth than at its center—even though there is proportionally little difference in the distances involved. (The distance between the surface of the Earth and its center amounts to only 1.67% of the distance to the Moon). An illustration of the Britannica's explanation for the second daily tide on the far-side of Earth appears in (Figure 2).

> "Thus it can be seen that the tidal forces tend to pull the water toward the Moon, (on the near side), and <u>away</u> from the Moon, (on the far-side)" [2]

---

1   Encyclopedia Britannica. 15th Edition. Volume 21, p. 1127, 1973.

2   Ibid. p. 1128.

THE DOUBLE PLANET

Figure -1-

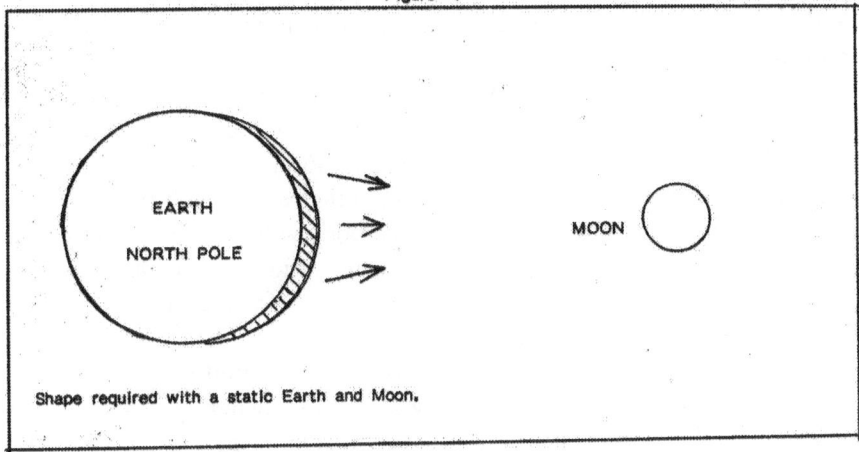

Shape required with a static Earth and Moon.

Figure -2-

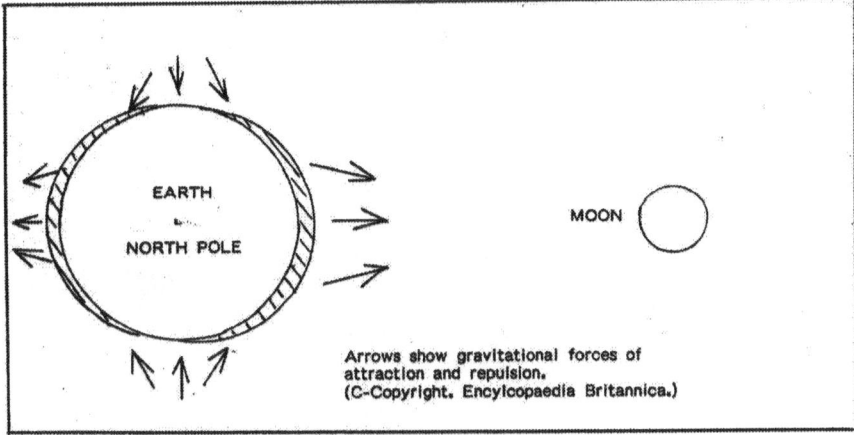

Arrows show gravitational forces of
attraction and repulsion.
(C-Copyright. Encylcopaedia Britannica.)

This explanation is unacceptable. Surely, the author cannot expect us to believe that the Moon exerts a repulsive effect upon the waters on the far side of Earth! This is nonsense. In reality, the second daily tide may be better explained in another way—by viewing an imaginary model of the Earth and its world-ocean as we would an experimental object.

If Earth was a totally liquid body, and if, with regard to each other, the Earth and Moon were stationary, the resulting form of the liquid Earth, *would* be egg-shaped; with the smaller end pointed toward the Moon. (Figure 1) But this is not the shape we actually observe. With the rising of the seas on both the near and far sides of Earth, what we see is an oblate-shaped body with its longest axis directed toward the Moon. (Figure 3) What creates this condition? If a single gravitational force would create an egg-shaped body, what other force could there be that would produce an oblate or lozenge shaped body?

Experimentally, such a lozenge shaped body may be produced by attaching a real, or imaginary, liquid-filled balloon by a line and swinging it about a pivot point. At a suitable speed, a lozenge shaped balloon will be created with an appearance quite similar to the shape observed in the hydrosphere of the Earth. (Figure 4)

THE DOUBLE PLANET

Figure -3-

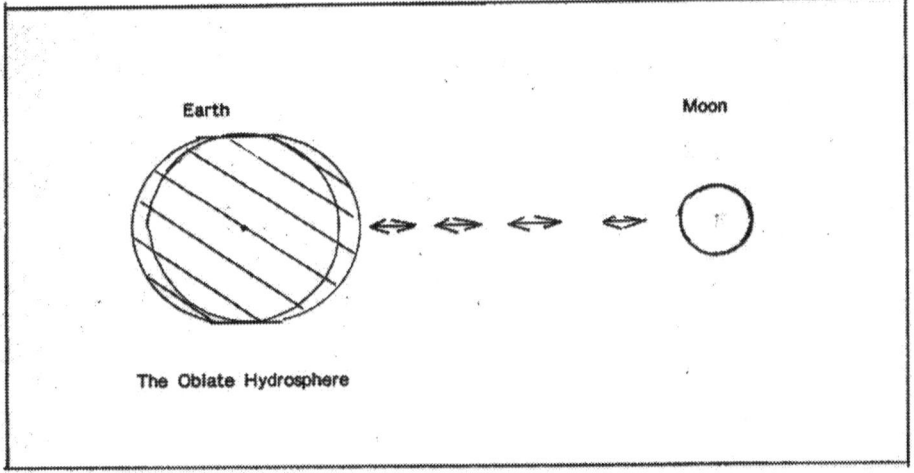

Figure -4-

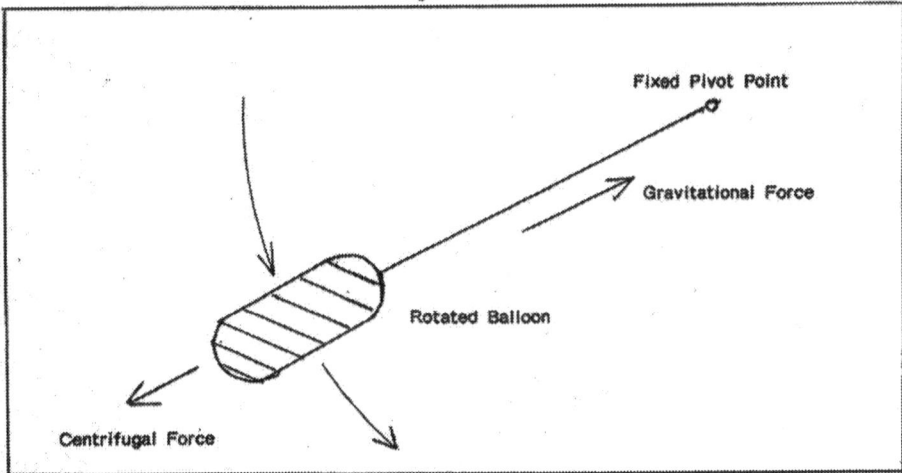

In this demonstration, the lozenge-shaped, liquid-filled, balloon results from a combination of two forces. First, the force of their mutual gravity, represented by the line attached to the pivot point and, second, by the centrifugal force created by the inward-turning moment applied to the balloon as it orbits about the pivot. This experiment duplicates the shape we observe in the Earth's oceans. But, in the relationship between the Earth and the Moon, where is that pivot point and what could be causing a similar centrifugal effect?

The answer is found by considering the relative motions of the Earth and Moon that, due to their masses and their distance from each other, form a double-planetary system. Neither is free because, being in close proximity, neither remains stationary with regard to the other. In relation to the Moon, Earth's position is not fixed and the Moon, likewise, does not orbit the Earth. Rather, both orbit about another point, their common center of mass. Their true motions may be illustrated by imagining a dumbbell-shaped object with ends of different sizes and weights rotating, freely, in space. In this illustration, we see that an unbalanced object does not revolve about either its larger or smaller end, or about some random point but that, instead, it revolve about its fulcrum point, about the center of mass of the entire system.

(Figure 5)

THE DOUBLE PLANET

Figure -5-

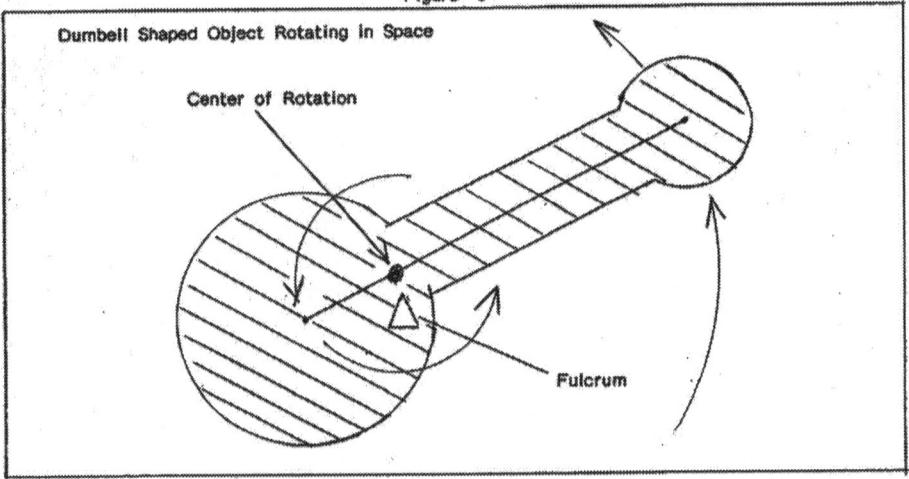

Figure 5 demonstrates that, if the rotating ends of the dumbbell-shaped object describe different sized circles, so, also, do the Earth and Moon as they endlessly pursue each other in their monthly cycles; the Moon, seemingly orbiting about the Earth, but with Earth, following after the Moon in its own, smaller, orbit; both rotating about their mutual center of mass. It is this motion of the Earth, within this smaller orbit, that creates the centrifugal force causing the second daily tide in the oceans. This second tide results from the centrifugal effect arising from Earth's motion as it follows the Moon in response to its gravitational attraction.

The fulcrum point of the Earth-Moon system may be calculated by comparing their individual masses and their present distance from one another other. Because the mass of the Earth is 81.3 times that of the Moon, and the distance between them is known, the fulcrum point is determined as follows:

THE MOON

Miles between the centers of Earth and Moon	238,855.00
Divided by the mass of the system	82.30
Multiplied by the mass of the Earth	81.30
Moon's distance from the fulcrum point	235,952.75

THE EARTH

Miles between the center of the Earth and Moon	238,855.00
Divided by the mass of the system	82.30
Multiplied by the mass of the Moon	1.00
Earth's center from the fulcrum point	2,902.25

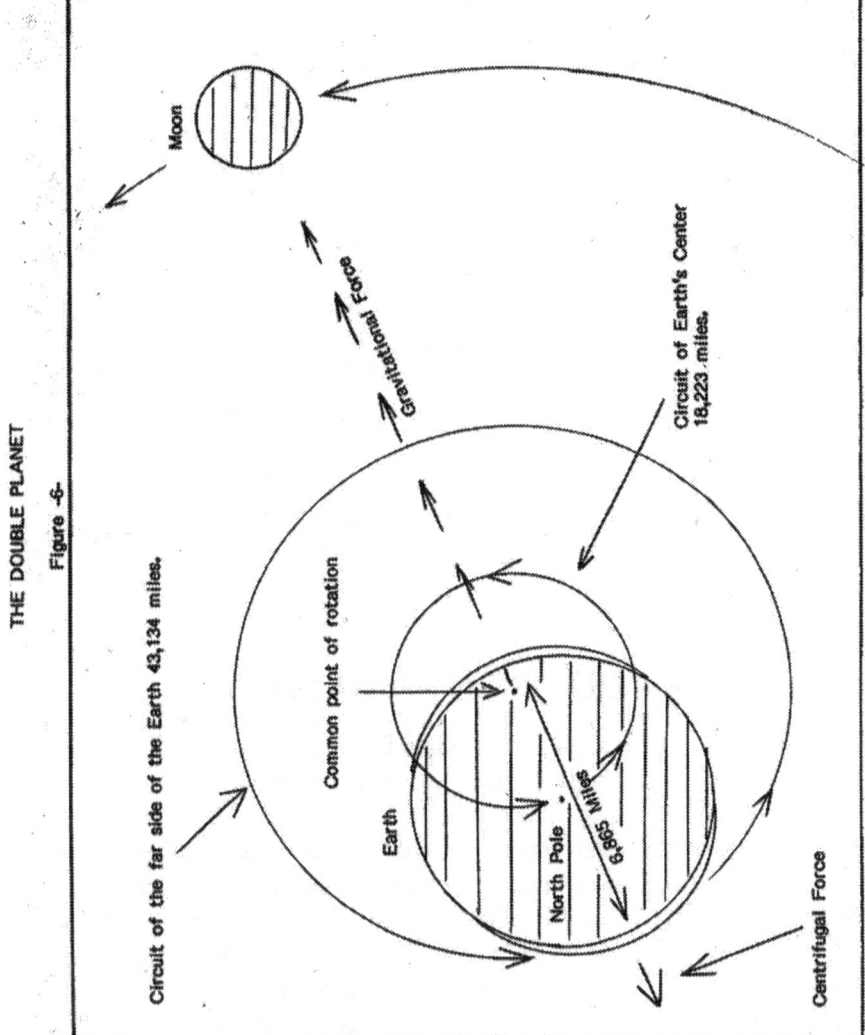

THE DOUBLE PLANET

Figure -6-

Moon

Circuit of the far side of the Earth 43,134 miles.

Common point of rotation

Gravitational Force

Circuit of Earth's Center 18,223 miles.

Earth

North Pole

6,865 Miles

Centrifugal Force

The center of mass of the Earth-Moon system is as shown in (figure 6). This illustrates that the fulcrum point, its center of rotation, lies 2,902 miles from the Earth's center and 1,061 miles below Earth's surface on the side facing the Moon at any given moment. The Earth-Moon system revolves about this fulcrum point once during each lunar month. Since the center of the Earth, is 2,902 miles outward from the fulcrum point, it is drawn about in a circular orbit, 18,223 miles in diameter covering an arc along this circle of 667.75 miles per day at an orbital speed of 27.81 miles per hour. The far-side of Earth, being 6,865 miles, outward from the center of rotation, circles in a larger orbit of 43,124 miles per month or 1,575.75 miles per day at a surface speed of 65.78 miles per hour. In accomplishing these motions, the far-side surface makes an inward arc of 14.18 degrees per day about the center of rotation. It is this centrifugal force, the force produced by this inward turning motion, that creates a reduction in gravity on Earth's farther side; the side away from the Moon. This lessening of the apparent gravity raises the second tide. The waters on Earth's far side are, actually, a liquid body, free in space, and as the Earth follows after the Moon, it, literally, withdraws itself from those far-side waters creating the apparent rise in sea level. The far-side tide that we assume to be a second lunar tide should more properly be called Earth's daily centrifugal tide.

THE PERMANENT WORLD TIDE

From another perspective, imagine viewing Earth from a point high over the North Pole. Exaggerated, the lunar and centrifugal tides would be visible as two bulges on opposite sides of the planet as in shown in (figure 7)

Over time, it would be seen these tides, that appear to us, as twice daily occurrences, are only illusions caused by a compounding motion; the, once daily, rotation of the Earth within these two tidal bulges. The tidal bulges would then be seen as, what they really are, permanent tides, created by the gravitational and centrifugal forces of the Earth-Moon system. From that height, we could see there is but one lunar tide per month accompanied by one centrifugal tide, with Earth rotating within these two permanent tides, once each day.

THE LUNITIDAL INTERVAL

Earth's rotation within the permanent tides invokes a further resultant. On Earth, the time at which the highest lunar tide is observed is not when the Moon is directly overhead, the time at which it ought to be expected but, rather, the high tide occurs considerably later. This time-delay is called the "Lunitidal Interval;" the period between the Moon's passage overhead and the next high tide. Were the

Earth a non-rotating sphere, high tide would, of course, occur exactly when the Moon passed overhead or, more correctly, when our location upon the Earth came to be on the direct Earth-Moon line. But, in point of fact, high tide occurs much later. In the latitudes of North America and Europe, the interval of delay is from 1-1/2 to 2 hours. Looking down once again from our polar observation post, we can see that the tidal bulges are displaced, Eastward, from the direct Earth-Moon line by as much as 30 degrees. This is due to the frictional effect of the Earth, as it rotates within the permanent tides, pulling their waters approximately 30 degrees Eastward. (Figure-8)

THE DOUBLE PLANET

Figure -7-

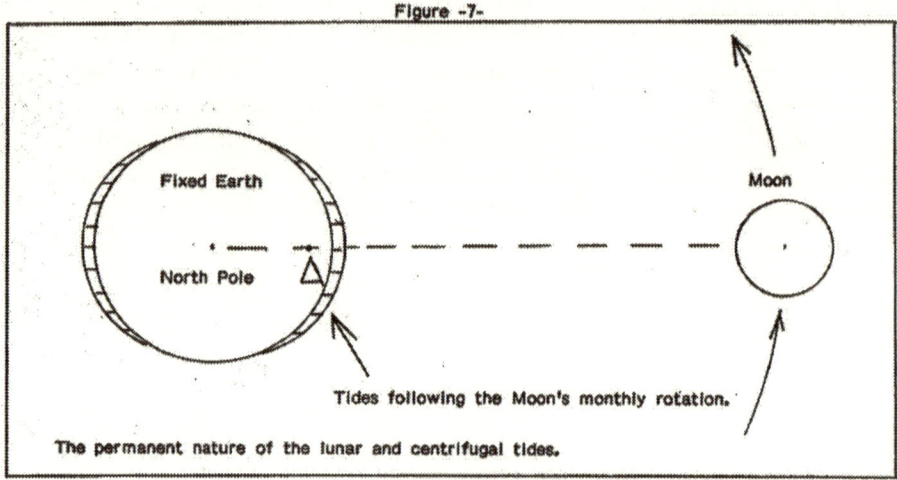

Fixed Earth

Moon

North Pole

Tides following the Moon's monthly rotation.

The permanent nature of the lunar and centrifugal tides.

Figure -8-

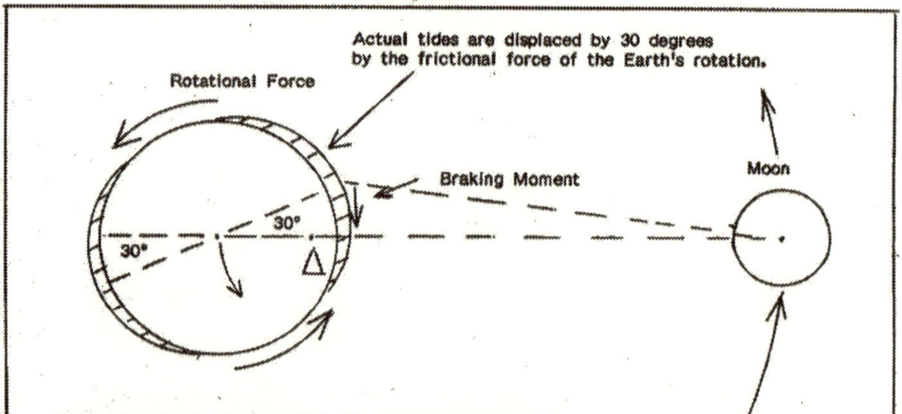

Actual tides are displaced by 30 degrees
by the frictional force of the Earth's rotation.

Rotational Force

Braking Moment

Moon

30°

30°

The actual Earth rotates within the permanent tides daily.
Frictional action displaces tides by 30 degrees and creates the lunitidal interval.

This displacement of the tidal bulges may be compared to the operation of a pair of automobile brake shoes that, while clamping down upon the outside surface of a rotating wheel slow the wheel, but as they do, must take up the energy of rotation and, themselves, be pulled part way round in the direction of rotation until they are held firmly by their mountings or, as in this case, by the gravitational attraction of the Moon. This illustrates how the Moon continually brakes the rotational speed of the Earth—even as Earth's, equal and opposite, rotational force accelerates the orbital speed of the Moon. These two forces, combined, equal the power necessary to raise and pull the weight of the tidal bulges, Eastward, a distance of 1,500 miles and perform this work continually. The strength of these forces has been estimated to be about 4 billion horsepower.

EARTH'S ROTATIONAL PERIODS

While it has long been understood that tidal braking might be reducing Earth's speed of rotation, in the past, there was never an accurate way to measure it. From the records of ancient eclipses, and from the quite accurate astronomical data kept since the year 1600, it is now known that Earth's annual rotation is "losing" at least 30 seconds per century. Prior to those observations, the slowing was undetected because the length of our day was psychologically "expandable." We naturally assumed the length of Earth's day to be fixed and we adjusted our means of timing to that standard. Now, with an absolute measure of time by means of the atomic clocks, we can directly measure whether there is any slowing. The atomic clock system was first established upon the base year, 1972, so that, Earth's period of rotation can now be compared against its speed in that year. Since then, it has been discovered that the Earth, today, is rotating at a slightly slower speed than in 1972. This slowing means that the Earth has "lost" about one second from its former time required to make its 365 rotations in a sidereal year. This loss of one second has necessitated the addition of annual "leap-seconds" to our atomic clocks to align them with the observed period of Earth's present rotational speed. Between 1972 and 2000, 23 of these leap seconds have had to be added. It is amusing to see that, even with the advantages of our modern means of measurement, we have not lost the convenient habit of "expanding time" by adjusting our clocks to Earth's day. This habit is difficult to break. It remains true, however, that rate of the Earth's rotation *is* slowing and the cause, undoubtedly, is the gravitational pull of the Moon.

THE EARTH'S ROTATIONAL SLOWING

As an equal and opposite reaction, to the Moon's slowing of the Earth's rotation, the Earth's rotation has, also, always been accelerating the Moon. It follows then that, in former times, the Moon's orbit must have must have been closer to Earth than it is now—and that Earth must, consequently, have rotated faster and had a shorter period of daily rotation.

"Some remarkable research by John W. Wells, of Princeton University, shows that tidal friction is not limited to the historical period of man but has been in progress since ancient geological periods as far back as the Devonian. He finds that certain fossil corals of this period exhibit annulations, like tree-rings, measuring their daily and annual growth rates. Thus the year (then) contained some 400 days, reducing the day to some 22 hours! C.T. Scrutton of Oxford University has since found monthly annulations, indicating the lunar month was, then, only 21 days! [3]

In a closer orbit, the gravitational attraction of the Moon must have produced far stronger oceanic and earth-tides than it does today. The greater earth-tides must, necessarily, have been the cause of the severe geologic diastrophism that crumpled the surface of the continents into mountain ranges while, at the same time, pulling the continents apart; redistributing them into their present locations. The frictional heating, caused by this daily kneading of Earth's crust melted the upper layers of the Mantle's rock and created Earth's volcanism. Gases from those volcanic eruptions, undoubtedly, produced Earth's early atmosphere. This illustrates the vital role of the Moon in providing Earth with the necessary energy to form a livable planet.

In his studies of thermodynamics, the famous 19th Century scientist, Lord Kelvin, calculated that, because he knew the rate at which the Earth radiates heat into space, he estimated that, even if the Earth had begun its existence in a molten state it would, by now, have lost most of its latent heat. From this he concluded the Earth could be no more than 6,000 years old! As one might expect, while his calculations were correct, they did not sit well with the messianic evolutionists of his time. But, what both he and his critics failed to realize was that, while Lord Kelvin's arithmetic was correct, another source of heat always was, and still is, continuously being re-supplied by the kneading of the earth-tides. This explains why Earth's average temperature remains steady at approximately 52.9 degrees Fahrenheit. It is because the heat that is steadily lost into space is being re-supplied by internal heat rising up from below.

---

3   Fred Whipple, "Earth, Moon and Planets," Harvard University Press, 1970. p.107.

Since Earth's period of rotation *is* slowing, and because our means of measuring time and distance have become quite precise, it should be possible, by mathematical analysis to extrapolate, backward, to determine the distance of the Moon from the Earth and the consequent periods of Earth's daily rotation, thousand or even millions of years ago. Because the gravitational forces between these two bodies, at lower and lower orbits, calculating backward, would not produce a straight line of regression but an increasingly larger function, in former times, the tidal effects of the Moon and speed of the Earth's rotation would have been far greater. One must ask why such a study has not been attempted? Perhaps, it is because it is feared that the results would show the age of the Earth cannot be "billions and billions" of years as required by the "Theory of Evolution." It might also show that, either the Earth is not as old as previously supposed or that the Moon may have come into its present orbit at a more recent date.

THE SEASONS

We experience our seasonal variations in climate because the inclination of the Earth's axis of rotation stands at 23 ½ degrees from the vertical, in relation to the ecliptic—the plane of Earth's orbit about the Sun. But was this Earth's angle of inclination from the beginning? It is evident, that all of the solar planets are related to one another in that they all orbit about the Sun in the same plane. If again, we could view the Solar System from a point high above the Sun's North Pole we would see that all the planets, also, circle the Sun, leftward in their orbits in a counter-clockwise direction. Likewise, all the planets, except Venus, rotate upon their axes in the same counter-clockwise manner. Venus, the one possible exception, is said to have a small "retrograde" motion with its day being over one Venusian year. This backward motion may be only superficially apparent and may be found to have another explanation. It follows that all the planets originally began with their axes of rotation at the vertical, 90 degrees to the ecliptic, and that their axes should have remained in this direction, forever, barring perturbing influences from outside forces. Thus the inclination of Earth's axis, at 23 ½ degrees from the vertical, indicates that some considerable external force must have caused it to tilt. Rotating planets are like gyroscopic bodies in space and their axes should remain fixed in direction unless an external force is applied. With gyroscopes upon Earth, this force is the effect of Earth's gravity. When that force is applied, the axis of a gyroscope begins to tilt and precess,—that is, its axis begins to slowly revolve in a circular path around its original orientation of 90 degrees from the horizontal—exactly what has happened to the axis of the Earth. Like a gyroscope, Earth's axis is presently tilted at an angle of 23 ½ degrees from the vertical and it precesses in a circular path, once in every 25,600 years, about its original orientation, 90 degrees. What caused this tilted axis and the resulting precession? The only possible answer is that the tilting of Earth's axis must be the result of the ongoing attraction of the Moon. The outer planets Saturn and Uranus, also have highly tilted axes and, in both cases, they too are accompanied by orbiting masses in the form of multiple moons and rings.

In the beginning, before the attraction of the Moon had created this tilting effect, Earth's poles *were* at 90 degrees to the ecliptic plane. This orientation created climatic conditions far different from those we experience today. Then, there was a uniform temperature with all parts of planet receiving sunlight 50% of each day—every day of the year. The climate was temperate at the poles, but was also temperate at the Equator because Earth, then, was under a deep and permanent cloud cover creating a "greenhouse effect." This greenhouse effect equalized the temperature over the whole planet. Fossil corals and other evidences of warm water life now to be found in the sediments of both the North and South Polar regions support this theory. The greenhouse effect was maintained because a large portion of Earth's water supply remained in suspension; floating overhead in the form of cloud vapor. The climate, then, was equable and the temperature had never fallen far below the dew point; and therefore, it had never rained. A quotation from the Book of Genesis provides us this description:

"(And God made the Earth and the Heaven) and every plant of the field before it grew, and every herb, in the field before it grew: for the Lord had not caused it to rain upon the earth and there was not a man to till the ground; but there went up a mist from the earth and watered the whole face of the ground. [4]

Whatever one's religious beliefs, it is undeniable that the Bible and other ancient traditions constitute the earliest record of man's historical knowledge. Because similar stories are recounted in the various ancient writings does not invalidate them. They are the account of a widely shared experience. The story of the Flood was given to us by those who, in their own time, had no doubt as to its reality. The biblical evidence carries further weight because, often, it has been supported by the archeological discoveries of more recent times. Recognizing that life, then, was lived under a "greenhouse" climate, explains why we find mastodon remains at all latitudes, in North America, Central America and in Asia. Mastodons, like modern elephants, required a temperate climate where they could find abundant grazing. They could not have lived in today's Arctic Zone. We know, also, that they were contemporaneous with our distant forbears. We see that earliest buildings of the Maya culture in Central America are covered with

---

4   Holy Bible, King James Translation. Genesis II, 5-6.

carved mastodon images—proving that these pre-diluvian people were well acquainted with them.

Mastodon remains are now more commonly found in North Polar latitudes where they are frozen in the tundra, but their skeletons appear at all latitudes; even in tropical Central America. John L. Stephens, the American archaeologist who, in the 1840s, pioneered the exploration of the Maya ruins in Central America, reported being shown the skeleton of a huge mastodon buried in a standing position in the side of a riverbank near Quezaltenango, Guatemala. The skeleton had just been unearthed by a recent flood. Stephens estimated it to have been from 25 to 30 feet long! [5] This proof of the existence of mastodon at these lower latitudes indicates that the world's temperature, in both Central America and in the Arctic was universally moderate rather than extremely frigid or extremely tropical as they are today. The climate had always been temperate since the beginning of life. The lands were covered with forests and the seas were filled with living creatures. Then, without warning, came a violent change in Earth's climate.

Speculative writers such as Vielevkosky and Erich Von Danikan have put forth theories that Earth's history has often been interrupted by cataclysmic events and it must be agreed that there once was a sudden change. But, these authors believe this change was the result of volcanic eruptions or the effects of a meteorite striking Earth, or from the close passage of another large planetary body.

But, does explaining such a change require the impact by a meteorite large enough to create, world wide, volcanism yet small enough so that it would not altogether destroy the Earth; a meteorite whose crater cannot be found today? Or, could this change more reasonably be explained by looking to that force which has always been operative and remains operative today—the gravitational pull of the Moon?

Making the reasonable assumption that the orientation of Earth's axis *was* originally at 90 degrees to the ecliptic plane, what could have altered Earth's tilt into its present position except the Moon?. It is the only force that could, logically, have caused this. The Moon's tilting of Earth's axis was, however, a gradual process and, initially, did not affect the climate on Earth. So long as the tilt was not extreme, the difference between the solar heating effects in the Northern and Southern hemispheres were not noticeable to the inhabitants and the deep cloud-cover damped out the solar heating variations. But, as the tilt approached the angle we observe today, (23 ½ degrees), a critical event occurred. The damping

---

5   John L. Stephens, "Incidents of Travel in Central America, Chiapas and Yucatan."
     Vol. II, p. 229. Reprinted by Dover, 1969.

effect of the cloud-cover was overcome by the further decrease in solar radiation being received by the "winter hemisphere." Because of this, in the "winter" hemisphere, the temperature of the clouds finally cooled to below the dew point—causing them to condense into rain and near the pole, into snow! Thus began that world-wide rain of cataclysmic proportions, a rain storm such as had never occurred before, and can never occur again; for it was the first rain—it was the Flood!

"... in the six hundredth year of Noah's life, in the second month, the seventeenth day of the month, the same day were all the fountains of the great deep broken up, and the windows of heaven were opened. And the rain was upon the Earth forty days and forty nights." [6]

Such a vast amount of water fell, and with such suddenness, that there was no time for it to percolate into the soil or to run off into the ocean basins in an orderly fashion. The whole surface of the land was flooded. As the waters deepened, the attraction of the Moon began to create gigantic twice-daily tides that swept around the Earth; devastating all life upon the planet. The fish and marine animals of the seas were flushed up and deposited upon the beaches and upon the lands—only to be buried by each succeeding tide under hundreds and thousands of feet of sediment. These tides swept up the forests and piled them in great reefs that were buried, in their turn, under succeeding layers of sand and silt. Yet, as destructive as the Great Flood was, its beneficial result for modern mankind is that it left us with the petroleum and coal deposits we mine and depend upon today. Fossil coal and oil are the chemical remains of the immense amount of marine, animal, and vegetable matter that was buried by the Flood. The geological record is proof. The coal and oil now found widely distributed are all of an equal age and come from a common origin. All are the result of that, single, hydrologic cataclysm. While the wholesale destruction of much of Earth's life was a disaster, what would our human situation be today had there not been a Flood? Without it there would be no coal, no oil, nor natural gas to power our modern civilization. Without these, our only source of fuel would be, what it was up to the beginning of the industrial revolution—wood or camel dung. Our means of transportation would still be by foot power, horseback or, at sea, by wind and sail. Underlying today's industrial civilizations, is the fact that it has only been made possible by the discovery and use of these fossil fuels.

After forty days and forty nights, the clouds broke open and the rain ceased. Then began the annual round of the seasons: Spring, Summer, Fall and Winter. Because the Sun now heated Earth's surface, directly, the equatorial areas became

---

6   The Holy Bible. Genesis, VII, 11, 12.

tropical and the polar regions, frigid. The temperature gradient between these two regions that, heretofore, had been negligible, now became extreme. The difference today is on the order of 120 degrees Fahrenheit. As modern weather began, chilled and denser air flowed down, along the surface from the poles towards the equator where it was heated and flowed back to the poles at higher altitudes. This climatic change overcame the mastodons that we now find frozen in the permafrost of Siberia. At the beginning of the Flood, the air in the Arctic suddenly chilled from +70 degrees to–50 degrees Fahrenheit within hours. On that day, the mastodons and every other living thing, above the 60th degree parallel of latitude, were quick-frozen where they stood.

For another 150 days, the waters ran over the land. Foaming rivers crashed downward toward the sea carving out the great canyons such as the Grand Canyon of the Colorado—creating them in a manner of weeks. The waters continually shifted; re-depositing the water-ground rocks and sediments that now cover Earth's surface. Proof of this vast churning and redistribution of rocks can be found by digging down with a pick and shovel anywhere on the planet. Invariably, what is found, just below the surface, are rounded, water-ground rocks of all descriptions, transported there from unknown locations; all mixed together with finer sand and gravel. Common sense brings one to the conclusion that this soil and these rounded stones must have been deposited by some vast hydrological force; not operative today.

"And God remembered Noah, and every living thing, and all the cattle that was with him in the Ark:: and God made a wind to pass over the Earth, (the first wind?) and, the waters assuaged. The fountains of the deep and the windows of heaven were stopped, and the rain from heaven was restrained. And the water returned from off the earth continually: and at the end of one hundred and fifty days the waters were abated." And God said: "I do set my bow in the cloud and it shall be for a token of a covenant between me and the earth" [7, 8]

It was the first rainbow and was the first time that the survivors saw, revealed, the Sun, the Moon and the Stars.

THE CONTINENTAL DRIFT

Geologists believe the land mass of Earth formed originally as a single continent floating over the denser molten Mantle. But, upon a rotating planet, this single continent created a state of imbalance that the earth tides began to correct. The

---

7   Ibid. Genesis IX, 11.

8   Ibid Genesis VII, 1,2,3,

Moon's kneading gradually pulled apart this single land mass and re-positioned the resulting continents to create a balanced globe.

"And unto Eber were born two sons: the name of one was Peleg; <u>for in his time was the earth divided,</u> and his brother's name was Joktan." 9

Reluctantly, science has come to agree with this biblical account. After 1960, geologists began to accept the theory of plate tectonics that asserted that the continents are like floating islands, continually in motion, floating slowly over the molten mantle. If, at the time of Peleg, the continents were still joined or separated by only narrow straits, colonization from one center of culture to another would have been possible. Is this the answer to many of the mysteries of archaeology that have puzzled researchers in the past? In Central America the Negroid features of the carved stones of the Olmec Culture show they had contact with Africa; and it seems that only the Olmec culture, the oldest to be found in Mexico, had this knowledge. All later cultures failed to depict persons or gods with Negroid features because their memories of Africa and the Africans had by then been forgotten.

How rapid was this continental separation? If at that time, the orbit of the Moon was considerably lower, its effects would have been proportionally greater. The parting of the continents would have been accompanied by violent earthquakes. It might have appeared to the peoples of Europe and Africa that their neighboring continent had just "disappeared!" They would have rowed out in their little boats to where this land had been and, finding it not there, would have concluded it must have sunk into the sea during the recent earthquakes when, in fact, it had only slipped back, over the horizon where they did not dare to go. Might this not explain the persistence, in myth and memory, of the story of Atlantis (i.e. America), the land that was described as being in the Atlantic Ocean, "beyond the Gates of Hercules?" (Gibraltar). Europeans called this mythical land "Atlan-tis" long before they knew of the American Continent. Yet we find the word and sound "atla" used today in many Aztec words. The Maya throwing spear was called an "atlatal and the word "atlan" still appears in Mexican city names such as "Maz-atlan." These people called themselves the "Atlans" and their principal god's name was "Quetzal-co-atl." Can the naming of the Western Ocean, by early Europeans, as the "Atlan-tic" be mere coincidence?

Could this series of violent movements of the Earth's crust not account for the otherwise inexplicable similarities between the architecture and artistic motifs of Pre-Columbian Central America and those of Early Egypt? The first Egyptian pyramids were step-pyramids that duplicated the style inherited from "Atlantis."

---

9   Ibid Genesis, X, 25.

The symbolic "all-seeing eye of Osirus," honored in ancient Egypt, is also to be found from an earlier time, carved upon the walls of a temple at Teotehaucan, near Mexico City.

A significant feature of Mayan buildings is that they are constructed as solid masses; beautifully faced with dressed and artistically carved stone on the outside; but, inside, they are solidly filled with rubble. They usually contain only a few small rooms comprising less than ten percent of their volume. The question becomes, why would they build in this manner? Why construct such large buildings with so little useful space inside? The explanation must be that, at that time, only massive buildings could remain standing during the persistent and violent earthquakes that were racking the Earth. If so, there was wisdom in their design—for while time and the climate have taken their toll, most of these structures still stand. Massive construction is the best solution for building earthquake-proof structures. Central America is still one of the most active earthquake areas in the world. During the last severe earthquake affecting Mexico City, the heavy, stone built, Spanish Colonial buildings, such as the Opera House and the Cathedral, survived intact—while modern buildings, built in the North American skyscraper-style, fell over or collapsed.

The pre-diluvian origin of the Ancient Mayan (Atlan) Civilization, in Yucatan, is confirmed by the area's present lack of a surface water supply. The Yucatan Peninsula is composed of a flat block of limestone covered with thin soils incapable of supporting a large agricultural population. It is a poor place. For the most part it is covered only by sharp edged cactus forests; inhabited by few animals but by hordes of voracious insects. There is no surface water. Even today, the only sources of fresh water are the underground rivers that run through caverns 100 feet below the surface. Without modern pumps, this water would not have been accessible. Yet, as we know, a large population once lived here. From the top of one of the renovated Mayan temples, one may look out over the jungle and see other mounds spaced one-half mile apart, marking innumerable other, as yet, unexplored sites. These mounds stretch on to the horizon indicating that, when they were built, there must have been a vast population with an efficient agricultural system capable of feeding it. The soils must have been much better and the climate more temperate. Quite likely, they collected their drinking water from the daily dew, exactly as described in the Book of Genesis. Archaeologists have now found what they believe to be their storage reservoirs which they have named "aguadas" carved out in the surface of the limestone rock.

The Atlantaens' monuments still testify to their technical skill and artistry— yet their whole civilization suddenly disappeared. Their works now lie abandoned in the dense tropical jungle because they, their soil, and their whole civilization was swept away in a single day—leaving only these ruins as their epitaph.

THE VOLCANIC EARTH

Earth has had a long history of volcanism. Yet the heat necessary to cause this volcanism could not have been generated and maintained without being renewed by some chemical, nuclear, or mechanical process. According to the laws of thermodynamics, as hot as the Earth may have originally been, its original heat would have been by brought to the surface by convection and lost into outer space. The theory that Earth's internal heat is being replenished by nuclear fission requires that a large amount of radioactive material be concentrated in a critical mass in order to produce atomic heating. The difficulty with this notion is that, if much of the interior is molten, the radioactive elements would dissolve within the melted magma—diluting their concentration below the critical point of fission. If we theorize that Earth's temperature is being maintained by the heat of the Sun, we face other problems. The temperature of the Earth's surface when exposed to constant sunlight, at Earth's distance from the Sun, can rise to 120 degrees Fahrenheit during the day, but, as we know also, it will then drop below freezing at night and the solar heat would not penetrate far below the surface. Earth's interior would be no warmer than its surface—yet as we know, it is far hotter.

The only remaining factor capable of creating Earth's internal heat is the "earth-tidal" force being constantly applied by the Moon's gravitational attraction. This force works upon the body of the planet; kneading the rock of the mantle like plastic dough. This kneading is the only force present capable of providing the prodigious frictional heat-energy necessary to melt rock within Earth's Mantle.

"Careful measurements of the tides in long pipes show that only 70 percent of the theoretical effects (of the Sun and Moon tides) actually occur. (in the seas).".... *"The main body of the Earth* yields *to the forces to the extent of the remaining 30 percent."* ... *"Outside the inner core, the Earth has an average rigidity of about twice that of steel."*.... *"A surprising additional result about the Earth was found from the tide experiments. The* Earth *is an* elastic *ball. The experiments showed the entire Earth yields* immediately *to the tide raising forces and it* immediately *returns to its original shape when they are removed.... The energy of the tides is thus changed into heat, at a rate of some 4 billion horsepower." Thus the tidal friction may take place in the crust as well as in the extended shallow waters.* [10]

The vertical distance of these movements may be only on the order of a few feet per daily cycle but, in a material as resistive as solid rock, they create an enor-

---

10   Fred A. Whipple, "Earth, Moon and Planets." Harvard University Press. 1969. p 97.

mous amount of frictional heat—heat that gradually, rises to the surface. Without this heating, Earth's internal temperature would steadily decline and become no warmer than its surface. Eventually this would lower the surface temperature down to the permafrost conditions of the Arctic zones. Earth would become a frozen and static planet. Mars provides an exact example of the climatic conditions to be expected upon a planet that does not have the benefit of a companion with the relative size of our Moon. After its formation, Mars quickly cooled to its present temperature of 100 degrees below zero Fahrenheit, exactly in accord with the thermodynamic calculations of a Lord Kelvin. Mars' temperature, now, is maintained by solar heating alone.

EARTH'S GEOMAGNETIC FIELD

Another unanswered question has been: "Why does Earth have a magnetic field? How is it generated?" How could this field be maintained in such a medium as the presumed molten, interior of the Earth? While the magnetic lines of force run roughly North, and South, they are not aligned with true geographic North or South but are offset (Westerly in the Northern Hemisphere and Easterly in the Southern Hemisphere). Further, the magnetic poles gradually shift their location so that maps need to be updated every few years to reflect their new positions. The magnetic poles are not to be found at 90 degrees North or South latitude along with the geographic poles but, rather, they are found to be at about 12 degrees from the poles at 78 degrees North and South latitudes. The North Magnetic Pole is now located in Northern Canada and it moves about 5 miles, each year, in a Northwesterly direction. The magnetic lines of force, at the magnetic poles, point straight downward indicating that the source of their magnetism lies deep within Earth's innermost regions. Theories have been advanced to explain how this magnetism can be generated—such as that the Earth contains a powerful permanent magnet—or that movements of the liquid core create magnetism—or, that there is within the Earth a natural magnet that continually creates the field. But all these seem implausible. Might there not be a simpler explanation that better fits the evidence?

From the study of seismic waves caused by earthquakes, it has been estimated that Earth consists of four concentric spheres, one within the other, separated by slip zones; allowing some movement between them. The outer sphere consists of the continental and oceanic plates that extend downward about 100 miles; to where there is a slip zone between it and the next sphere, the Mantle. The Mantle then extends down to a depth of 2000 miles. Within the Mantle, is a, semi-fluidic, Inner Sphere of mixed iron and rock that extends, downward another 1,050 miles. Finally, within this Inner Sphere is a solid iron core. This solid core is

1,700 miles in diameter, with a radius of 850 miles from its center. It remains solid because, at this depth, its temperature is much cooler than that of the upper layers. The highest temperatures are to be found in Mantle because it is the most affected by the Moon's daily kneading. Below the Mantle, the temperatures may decreas as one probes deeper.

"Some 800 miles from the center, an inner core of very high density has recently been discovered. It may be solid." [11]

It is from this Inner Core, this sphere of iron, 1,700 miles in diameter that the lines of magnetic lines of force are emanating. Given this, how is the magnetic field generated and maintained? Is this sphere a permanent magnet? If it is, how could the field maintain its strength? And, finally, why are the magnetic poles not located at the geographic poles?

These questions can be answered by returning to our earlier assumption that, at the beginning, the axis of the Earth's rotation *was* oriented at 90 degrees to the ecliptic and that all the materiel contained in the planet must have shared in this original motion. But, later, as the gravitational attraction of the Moon began to exert its effect, the Earth's axis began to tilt towards its present angle. But, while this was the case with the outer layers, the solid, innermost core, floating within the semi-fluidic Inner Sphere, did not fully comply with this change. While the rotational axis of Earth's outer layers are now inclined at 23 ½ degrees from the vertical, the Inner Core followed by only one-half that amount; 12 degrees from the vertical; and because this Solid Iron Core is a body, free in space, it is partially exempt from the gravitational braking of the Moon; therefore it revolves at a slightly faster speed than that of the outer enclosing layers. This differential in their speeds of rotation creates a frictional "Dynamo Effect." The difference in the speeds of rotation between the outer layers and the Solid Iron Core can be explained by understanding that, while the Moon is braking the rotation of the outer layers, the Solid Iron Core is the last to feel those effects; thus there is slippage. It is this slippage, this differential motion between the Inner Sphere and the Solid Iron Core that generates the magnetic field. It should be observed that, while the Solid Iron Core now precesses backward (Northwestward) against the rotation of the Earth, it still precesses about its original orientation of 90 degrees to the ecliptic. (See Figure 9)

---

11   Ibid p. 84.

THE DOUBLE PLANET

Figure -9-

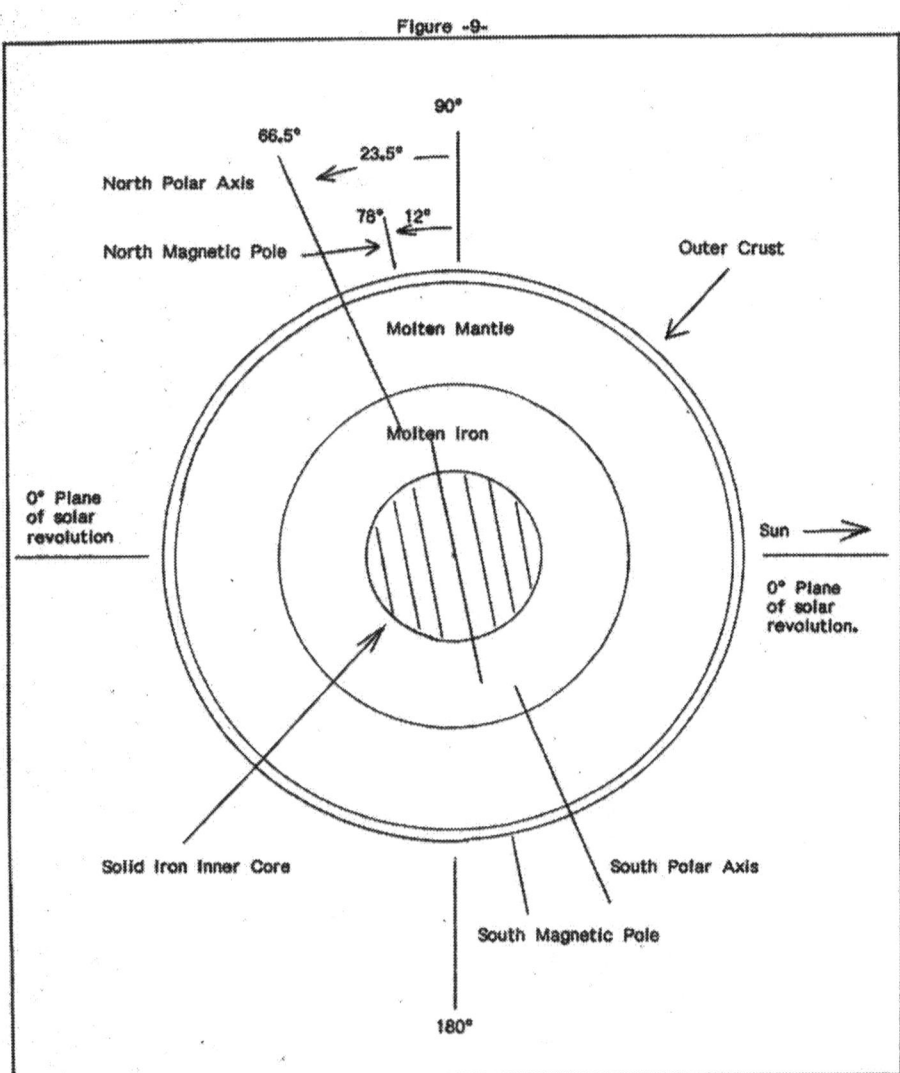

This conclusion is supported by the discovery of a similar situation on the planet Uranus. When Uranus was visited by the Voyager II space craft, it was found to have Saturn-like rings and five larger and ten smaller moons. Their combined gravitational attractions have tilted the axis of that great planet to an amazing 98 degrees from the vertical! The Northern Pole of Uranus now points 8 degrees *below* the ecliptic. But, as in the case of the Earth, its Solid Iron Core has only complied by one-half this change and Uranus' magnetic poles are tilted by only 43 degrees from the vertical. (See Figure 10) The differential speeds of rotation between the planet's outer layers and its solid Iron Core generates Uranus' strong magnetic field in the same manner as with the Earth.

THE DOUBLE PLANET

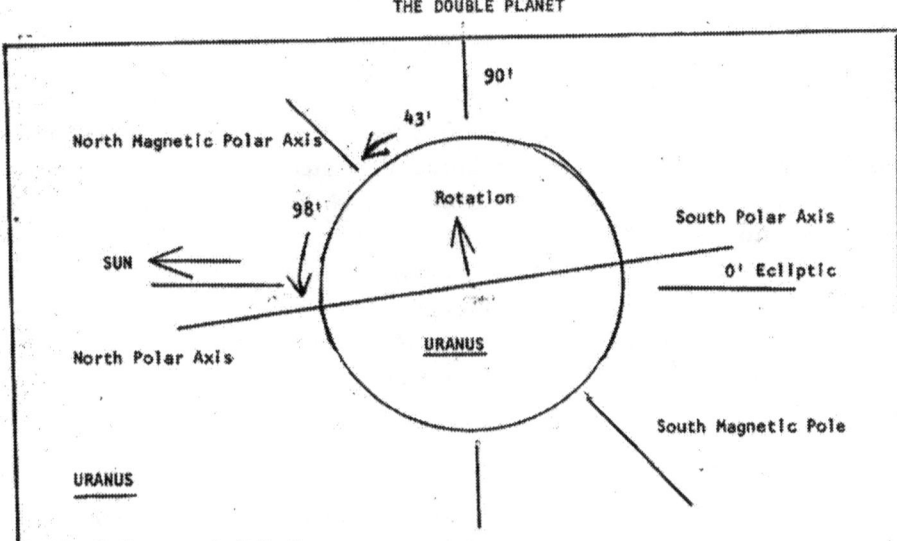

Figure -10-

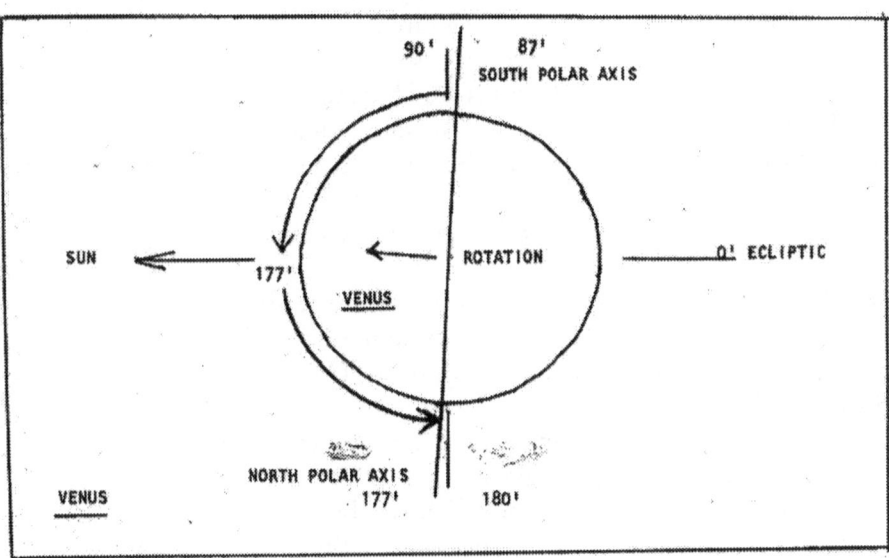

Figure -11-

Another surprise was that, since it is located at 19 times Earth's distance from the Sun, Uranus was expected to be a frozen planet with a temperature of minus 300 degrees Fahrenheit. But, when on its fly-by, Voyager II measured its temperature it reported that beneath the clouds, Uranus appears to be covered by an ocean of hot water or liquid methane 5,000 feet deep! This surprising condition must be the result of the gravitational kneading created by the attractions of its 5 major moons. This presumed kneading must continuously replace the heat energy lost into space by radiation. Another surprise was that of the temperature of Uranus' South Pole. Since that pole now faces away from the Sun during this part of its orbit, it was expected the South polar area would be far cooler than the, opposite, Northern pole; but this was not true. The temperature at both poles was found to be nearly identical due to the "greenhouse" conditions provided by the planet's deep and permanent cloud cover.

As with Uranus, Earth's double-planetary relationship with our Moon provides the same essential energy needed to maintain our temperate environment. Within this environment has come into being the greatest of all miracles, organic life. No philosophy so far has explained the power of the "life-force" or its ability to maintain itself. The "life force" is the very antithesis entropy; the negative theory that argues that all activities in the physical universe are gradually running down. Instead, the "life-force" continues to deny entropy as it sustains and renews every living form; plant, animal and human. That it does so is an ongoing miracle; totally beyond our powers to explain. Ultimately, it must be accepted as the providence of God or whatever else one wishes to call "the great mystery of life."

Officially, our scientists are not supposed to entertain any religious thoughts but, in truth, they think of little else. Our space probes are not sent outward toward the other planets to discover a better source of fuel or possible sites for colonization—no, our space scientists frankly admit that their real goal is to discover "organic life" somewhere else in the Universe. Their quest is, fundamentally, a religious undertaking.

If the above is true, what can we do to head off the danger of global warming or for that matter the danger of global cooling? The answer is, we cannot control them at all because the temperature of Earth is being maintained by a force that is out of our control; by the gravitational attraction between the Earth and the Moon.

What we can and should do is to limit the burning of fossil fuels by switching to nuclear energy as soon as possible. The most important goal should be to maintain our atmosphere in a healthy condition by carrying out a world-wide program of reforestation to maintain the oxygen/carbon dioxide balance and to reclaim Earth's desert areas for productive uses.

It is the atmosphere we must conserve and maintain because, realistically, we are all like the fish. We live in an ocean of air composed of 79% nitrogen and 21% oxygen and, like the fish, we cannot survive out of our element. Conserving the quality of the atmosphere should be our first priority. The imaginary fears about global warming or global cooling will, essentially, take care of themselves.

EPILOGUE
WHENCE THE MOON?
WHAT IF …?

Imagine that, somewhere, long ago and far away, there was a world similar to our own, inhabited by a highly developed race of beings whose scientists informed them that their planet faced certain destruction? Perhaps the doom foreseen was a cloud of asteroids that would eventually pass though their orbit and bombard them unmercifully? Perhaps it was the threat of increasing radiation from their star. Technically far more advanced than we are today, these creatures had invented a means of space travel making it possible for them to explore other the planets in their planetary system. They knew that a planet near them might, after the general disaster, be livable and they knew they faced certain extinction if they remained where they were. Therefore, they determined to build a space ship and evacuate themselves and as many life forms as possible; moving their culture to a new home.

To build such a ship was an enormous task. It had to be large enough to contain all the life forms of their planet, the fuel and supplies necessary for a long voyage and to enable them to set up their culture on the new planet. It would need to be immensely strong; able to resist the impacts of the meteor storms they knew they would encounter. With no other choice, they made a desperate effort to survive by building a monstrous ship in the form of a gigantic metal sphere. This they assembled, piece by piece, in orbit above their home planet.

When completed, the ship was loaded with all that could be carried away. Then, they set out. As their means of propulsion they used what we would describe as an anti-gravity beam, similar, perhaps, to a gigantic laser in our terms. However, for this beam to work it had to be directed against some other solid object to produce its repulsive effect. To create the maximum force, this beam was directed against the advancing edge of their own rotating planet to, in effect, to sap away that planet's rotational energy in order to drive their ship, out along the plane of the ecliptic toward their new home. So massive was this ship and so powerful was the repulsive beam that, by the time the ship had reached escape velocity, and was finally on its way, it had absorbed all of that planet's rotational speed and stopped it dead. This tilted the planet's axis into a new alignment tumbling it nearly upside down! When launching our own spacecraft, we, too, employ a similar method by directing them first Eastward; thus, gaining an extra one thousand miler per hour of orbital speed by using up some of the rotational energy of our own Earth.

Their journey was long because it was necessary, on the way, to survive in space for perhaps years, decades or even centuries during the holocaust of meteor bombardment and solar radiation they had to endure. The meteors could have been caused by the break-up of the unnamed planet we know must have orbited our Sun between Mars and Jupiter; the remains of which continue to orbit the Sun as the Asteroids.

Finally, when the bombardment had ended, they were able to approach the new planet. Using the same means of propulsion, in reverse, they slowed their ship and placed it into orbit. Their anti-gravity beam produced a similar effect upon the new planet, slowing its rotational speed, somewhat, and causing its axis to tilt from the vertical. But, because this planet was larger, the result of this tilting was not so extreme. They then descended by smaller craft and began to explore their new home; making for themselves a new beginning.

Is this story fantasy, or could it be true? If true, could the old planet be what we now call Venus and could the new planet be our own Earth? As fantastic as this account may seem, it fits well with the physical facts. But, where are those Venusians now? Do we need to ask? If the story is true, the Venusians are us!

Where is their great ship now? The answer may be that we see it almost every night but we call it our Moon—our companion planet that contributes so much to our life upon the Earth. Consider the evidence:

First: Certainly Venus must have, originally, rotated with its axis vertical to the ecliptic, counter-clockwise, as do all the other planets. This must be accepted as an absolute certainty. How could there have been an exception, in the direction of rotation of any one planet? In the scientific literature, Venus is described as now having a retrograde (i.e. backward) direction of rotation with its poles inclined only 3 degrees from the vertical. But on this point the experts may be wrong. The slight remaining rotational motion of Venus is still counter-clockwise, as are the motions of all the other planets but the North Pole of Venus has somehow been tumbled so far over that it is now at minus (-177) degrees from the vertical and is pointing almost straight down! (See Figure-11) In the past, it was believed it had to have a retrograde motion because, heretofore, no one could imagine that Venus could have been tilted over to this degree—but it is. What other imaginable force could there have been which could have sapped away the rotational energy of a whole planet, without destroying it, if it were not the space ship's repulsive beam?

Second: A perplexing connection between Venus and Earth noted by astronomers is that, each time Venus and Earth are in opposition, that is, at their nearest point to one another, the same side of Venus is turned toward the Earth. The cause of this phenomenon might be that the space ship was necessarily launched when Venus and Earth were in opposition and nearest to each other and thus, this was the final position of Venus when the space ship departed for Earth and its repulsion beam was turned off.

Third: Of ongoing concern has been the question of why the density of the Moon does not correspond with its volume in the same ratio as do Earth and the other inner planets. While the Moon has a cubic volume of X .02030 that of Earth, its mass is only X .01229 that of Earth—approximately one-half of what should be expected. The explanation may be that the reason the Moon has a lower density than Earth is because it is not solid at all—but a hollow sphere—a space ship; standing in orbit. While we note rocks and dust upon its surface, these may be merely the remnants of the meteor storms that, in the past, impacted upon its surface. In the smoother areas that we call the "Maria," (or oceans) these, perhaps, are the surface of the ship—its metal hull; areas that have remained uncovered or have been cleared of rocks and dust by subsequent impacts.

Fourth, and finally, knowing that no written or physical monument of their accomplishments would survive the ages, they placed their ship into an orbit at the exact distance from Earth that would provide all future inhabitants of Earth with the occasional spectacle of an annular solar eclipse. At its present distance, at 238,856 miles from Earth, the Moon, which is 2,106.6 miles in diameter, exactly subtends the disk of the Sun; making an annular total eclipse upon Earth, just possible. This positioning of the Moon, into an orbit, at that precise distance from Earth, cannot be the result of any celestial accident. This placement, at that exact distance and upon the ecliptic, must have been the deliberate act of a race, wiser than ourselves, who knew that, eventually, we would rise to a level of understanding whereby we could read and recognize their sign; the sign that they had succeeded in their quest to find for themselves, and for their descendents, a new Heaven and a new Earth.

Is this story fact or fantasy? Future missions to the Moon may tell us. For now, it remains an open question—What if?

Les Crane © 2007

"THE LIVES OF THE ELECTRONS"

ACTION AT A DISTANCE

©LESLIE C. CRANE

2007

Currently, there are deep concerns in the scientific community as to whether there can be such a thing as "Action at a Distance." Science has busied itself for the last three hundred years with the study and description of mechanically and chemically connected events; using as its basic hypothesis that physical or chemical force, "A," acting upon object "B," mechanically or chemically creates the resultant, "C." While this approach has utility, it is a convenient fiction because it fails to describe the most basic forces to be found everywhere in the universe such as electromagnetism, light and gravity. These forces demonstrate, "Action at a Distance." In truth, wherever we look, we see "Action at a Distance" in operation.

"Action at a Distance" is demonstrated by the repulsive force that exists between two magnets of like signs. We know it is literally impossible to force these two magnet poles together. Observing this, the question must be asked: "What is there, physically, between them that keeps them apart"? The answer seems to be that *nothing*; nothing, whatsoever, is between them. Yet, as we know, the force of repulsion is quite real. In the opposite case, the attraction between two magnets of unlike signs is equally mysterious. What is the nature of these attractive and repulsive forces? That is a fundamental question. Its answer might reveal the underlying structure of the universe.

ELECTRICITY

To describe electricity, in common terms, imagine that every electron in the atoms of a copper wire demands it be spaced at a predetermined distance from every other copper electron. However, when a loop of wire is rotated between permanent magnets as within the armature of an electrical generator, the electrons in the atoms of the copper wire "sense" the existence of the electrons in the magnets so that, intermittently, there appears to be an excess of electrons in their immediate vicinity. In order to reestablish their correct intervals, the electrons retreat; displacing themselves laterally along the wire. This shift of positions passes along the wire at about three-quarters of the speed of light. The electrons, themselves, do not move at this speed, but rather, what proceeds along the wire at this speed is the "readjustment" of their intervals. The electrons, individually, move only a short distance. An analogy might be to imagine the commonly seen row of standing bowling balls resting together in a rack. When struck by a moving ball at one end of the row, a single ball at the farther end of the row is, immediately, driven off at the same speed as that of the first moving ball; yet the intervening balls are almost unaffected. That every electron *knows* it must remain spaced at a certain fixed distance from every other electron is the basic principle underlying the phenomenon we call "electricity." This same expression of "action at a distance" describes electromagnetic induction, condensers, transformers, and all the other devices used to compress or to drive electrons in metallic circuits.

Unknowingly, what the early experimenters had discovered was a basic principle of atomic structure; that electrons, *know* their spatial positions vis-a-vis each other and that they *must* maintain these intervals at all costs; but, also, that they may be made to respond within an electrical circuit by the application of an external magnetic force. Electricity *is* Action at a Distance—the transfer of "knowledge" from one electron to another.

RADIO

During the earliest demonstrations of the radio effect, an alternating current oscillator was used to charge up a metal ring containing a small gap. When this oscillating current was applied to the ring, it caused an arc to jump across that gap. If another ring, having a similar gap, was set up at some distance in the same room, a smaller arc occurred across the gap in this second ring as well. It seemed obvious to the experimenters that the energy for the second ring's spark must have come from that first ring, i.e., from the transmitting ring, across the room to the receiving ring. At that time, the existence of the second ring's spark was explained by assuming the first ring's energy was being transferred to it through something posited as the "ether," because, it seemed obvious to them that energy could not, by itself, leap across empty space. Since then, the "ether" has been proven not to exist but we still know that something *is* being transmitted across empty space—but what is it? In the last 125 years since the discovery of the radio effect, our transmitters and receivers have been so improved they may be set further and further apart and are now able to detect extremely weak signals between each other; but the underlying principle remains the same. Information *is* transmitted and received while nothing whatsoever travels between the sending and receiving apparatus. As in the case of electricity and electrical circuits, all that travels between them is "knowledge:" The electrons of the receiving antenna "recognize" that, at a distance, there is a set of vibrating electrons in a transmitting antenna that is "attuned" to its frequency. The electrons in the receiver take up these vibrations, in harmony with those of the transmitter while nothing at all, physically, passes between them. The ability of electrons to communicate with each other over vast distances is limited only by the strength of the transmitters and the sensitivity of the receivers. The electrons in the antenna of a radio telescope recognize and harmonize with the vibrating electrons of stars that are light years distant. While at the other extreme, we have received and understood signals from the space probe, Mariner II, as it passed by the planet Saturn with a radio transmitter having only the strength of one watt!

LIGHT

The phenomena we call "light" is our ability to recognize that portion of the electromagnetic spectrum with which the electrons of our visual senses, are able to harmonize. The whole spectrum of electromagnetic vibrations is alike in character. Radio frequencies are lower than those our eyes can apprehend, while, ultraviolet, and "X-rays" are far above our visual abilities.

An ongoing paradox still troubles physics as to what *is* the basic nature of light? Is it made up of waves of "energy" or does it arrive at our senses as discrete packets called quanta or photons? Does light have mass? Heretofore, these questions have never been resolved because; depending upon the situation and the devices used to test these hypotheses, light appears to be in one case wavelike and in another quanta-like. This paradox continues because of a misunderstanding as to the basic nature of what is being observed. Heretofore, science has been unable to free itself from the notion that some physical material comes to us as light and, thus, we have necessarily deemed the pulsations of light to be waves within a medium or bundles of matter, intermittently, flying in our direction from the source. But, are either of these ideas true? If so, why have we not been able to prove the material nature of light during four centuries of experimentation? Might the solution not be that, because light is a portion of the electromagnetic spectrum, "light" is *not* material but, like "induction" or "radio" it is "vibrational information" mutually shared by the electrons of both the transmitter and the receiver—by the electrons in the Sun with the electrons in our eyes?

Imagine standing in the open air with our backs to the Sun but facing a yellow painted wall to our front. The electrons of the sun are vibrating in a wide range of frequencies from the infrared through the visible and on into the ultraviolet. The electrons on our world, including those in our bodies and in the wall before us, "know" the electrons of the Sun are vibrating and they vibrate, in harmony, but only at the frequencies with which they are, individually, attuned. On the backs of our bodies, our skin electrons recognize the frequencies of the infrared and as, they vibrate at an increased rate, we feel the sensation of heat. To our front, the electrons in the wall's painted surface recognize and reverberate with another solar frequency. Because of its composition, the electrons in the paint are attuned more strongly to the frequency we interpret as "yellow" and, because the solar electronic vibrations do contain this frequency, the electrons in the painted surface take up this "yellow" frequency as well.

Then, because the electrons in the retinas of our eyes are also responsive to the frequency, "yellow," they mimic the wall's vibrations. In our eyes, those vibrations create electrochemical, battery-like, currents in the rods and cones of our retinas and these vibrations pass the, "yellow." electronic vibrations, along our optic nerves into our brains, where this "information" is interpreted as the brain's mental picture of the external world with a yellow wall appearing in plain view. It is no exaggeration to say that, as we view an object, the "light" we see is not a substance reflected back from that object, like tennis balls bouncing back from a flat surface but, rather, it is the handing over of "knowledge" passed on to us from one set of electrons to another. Light is, essentially, "information," generated by the harmonic vibrations of the electrons; from within the object itself. Light is another demonstration of "Action at a Distance."

A PARADOX RESOLVED

Normally, "light information" is randomly polarized, That means the information we receive from vibrating electrons from the Sun, and from other sources, vibrate in all possible planes. Polarizing filters block out most of these planes so that a single plane of information, *polarized light,* passes through the filter. In a classic experiment, Dr. Abner Simony of Boston University, made a study of light polarities and in his article published in <u>Scientific American Magazine,</u> in the January 1988 issue, entitled "The Reality of the Quantum World," he reported he had discovered a certain paradox resulting from his experiments. His work involved the measurement and comparison of the polarities of individual, so-called, "light-photons." After exhaustive tests, with the most advanced equipment, he found that two photons, which initially had varying polarities, when arriving at the test equipment, somehow reoriented themselves within the apparatus, and exhibited matched polarities when they exited. This surprising result, in the words of Dr. Simony, indicated that the two photons, somehow, had the ability to compare their planes of vibration and to realign these planes with each other. This led Dr. Simony to conclude that this is an example of "action at a distance" a thing declared to be impossible by the Theory of Relativity and by the laws of quantum mechanics.

The seriousness of Dr. Simony's results provoked a firestorm of protest from certain "scientifically correct" scientists who argued these results cannot be true because they violate the "laws" of quantum mechanics and the "laws" of relativity. Further, if these findings were to be true, they ought to be suppressed as heretical! What follows is a letter to the magazine's editor signed by Professors, Oreste Piccioni, Werner Mehlhop and Brian Wright, of the Department of Physics at the University of California at San Diego. It appeared in the July, 1988 issue of <u>Scientific American.</u>

"In the "Reality of the Quantum World," Abner Simony appears to assess the Rinstein-Podolsky-Rosen paradox, (R.P.R.), as a very important issue. We agree. (The most typical example of RPR is the proposition that two photons, distant from each

other, and independent, un-polarized, show the same polarization along any arbitrary x or y axis when both of them are measured.)

We do not see, however, that quantum mechanics, in the context of R.P.R., or anywhere else, strongly suggests the existence of actions at a distance (AAAD) which is what explaining the R.P.R. in terms of "non-locality" essentially amounts to. Such a fantastic phenomenon would be nothing short of a miracle, and it would push our science back hundreds of years! Among other unacceptable features, it would violate the Theory of Relativity, because it would imply that unpredictable information produced in one place, would have a clear effect at another, very distant place, with negligible delay."

Yes, perhaps, it is a miracle. But are these results more miraculous than the Actions, at a Distance, of electrons in the transmission of electrical and radio signals; examples of which have been available for interpretation for over a century? The answer to the R.P.R. paradox is that it was not "photons" that were being measured as they exited Dr. Shimoney's equipment but pure "information" which, as it arrived at the test equipment as differently polarized information from two separate sources, merely energized the electrons of the first layer of the glass in the test apparatus and were then handed off, electron by electron through the apparatus; emerging, finally, as identically polarized information. This was because the electrons of the apparatus, associated with each other in a crystalline structure *were* polarized in phase with each other. The observers did not detect "photons" entering and passing through the apparatus, they detected "information" given them by the crystallized glass electrons at the output side of the lens. The paradox as to why light "photons" appear to be both wavelike and quantum-like, depending upon the equipment used to analyze them, is resolved by understanding that light is "informational" and not material. Such experiments indicate that "Action at a Distance" is the normal state of the universe. The physical universe, ultimately, consists of "information" being passed continuously from one set of electrons to another. Poetically, it is "The Music of the Spheres."

NEWTON'S RINGS

When we accept light as nonmaterial, we come to a solution of the problem of Newton's Rings. These are the black and white rings formed on a receiving surface by a beam of sunlight passing through a small pin hole; or about the edge of the shadows of small objects or in thin films, soap bubbles or oil spots on a wet pavement. They are of common occurrence but have baffled scientific observers for centuries. The fundamental question has been *why*, when the total surface area appears to be equally exposed to a light source, are there areas, at the edges, lighter and others darker? However, when "light" is viewed as pure information, the explanation becomes readily apparent. What actually occurs in this situation is that, in the areas making up the darker bands, the surface electrons are receiving, not too little information, but too much! Because the vibrational information between the electrons moves through space at a fixed speed, and because some of the information in passing through a pinhole is diffracted (i.e., it travels a slightly longer distance) it appears at the receiving surface slightly later than the direct vibrational information. The darker bands result from the addition of two sets of information exactly out of phase with each other. Since they are receiving contradictory information, the electrons in that portion of the surface do not sympathetically vibrate. They remain at rest and we recognize them as motionless and dark. The brighter areas are those exposed to a single harmonic frequency and therefore they *do* vibrate and are recognized by our eyes as lighted. A similar phenomenon occurs within an oil drop on a wet pavement or within in a glass prism. These exhibit, to our eyes, the colors of a rainbow spectrum. When vibrational information is received by a thin film (which does have some thickness) or through a triangular prism, vibrational information is passed through the film or prism, electron by electron. However, because the oil or the glass slows the vibrational frequency slightly and because of the varying distances they travel within the medium, the information passed on to the receiving surface, and thence to our eyes, causes us to see colors that are, in fact, only variations in the frequency vibrations in the information entering the receiving surface.

THE "SOLAR WIND"

Within conventional science, one of the "proofs" that light must have material existence is the concept that, in space, light from the Sun exerts a physical force upon on freely moving matter. However, experiments to replicate this physical force of light in the laboratory have led to inconclusive results.

Early attempts to measure the pressure of light led to the invention of the radiometer by Sir William Crookes. In this instrument, freely moving vanes on a pinwheel were coated with dull black paint on one side and polished on the other side. The black sides were expected to absorb radiation while the reflecting sides would reflect it back; thus there would be more energy (a greater force of light) at the front of the polished reflecting sides. However, when the radiometer was illuminated by sunlight, it turned, but in the wrong direction! The explanation given for this effect by Sir Richard and by other conventional scientists is that

"This is the result of a rather complicated effect depending on the heating of the residual gas in the vessel, and in order to observe the pressure of radiation much more delicate means are required." [12]

---

12   Encyclopedia Britannica, 15th Edition, 1973, Vol. 27, p. 1110.

Here is an amusing illustration of an experiment which, when it failed to produce the desired result, was re-jiggered until it did. The correct explanation of this perplexing phenomenon is that the electrons on surface of the black painted sides of the instrument's vanes, are more activated by the light "information" than are the electron on the polished sides of the vanes and thus they emit back more vibrational, heat, information.

Another asserted proof of the pressure of light, is the concept of the "solar wind." Its existence is "proved" by observing that interplanetary comets always have their tails pointed exactly away from the direction of the Sun. This occurs without regard to the comet's relative motion. This perception requires that we assume that "light" has mass, a fact never scientifically demonstrated. In reality, the behavior of these comets' tails may correctly be understood by, again, accepting light as information only. What actually occurs is that, as a comet approaches an active body such as the Sun, the free electrons in the comet's gasses recognize the Sun's electronic vibrations and, harmonically, vibrate in synchrony; thereby becoming "visible." At the same time these molecules of gas in the comet's tail, by induction, take up identical positive or negative electrical charges and, because like charges repel, the tail of the comet flows outward; away from the Sun's direction. This demonstrates once more "Action at a Distance," as the charges of the electrons in the Sun, and those in the comet's tail, recognize and repel each other over distances of million of miles.

GRAVITY

Sir Issac Newton's recognition of the gravitational force between masses, and his analysis of them by the use of mathematics, was one of the greatest advancements in the history of science. But, what *is* "gravity?" One important fact is that it is universal. Its effects are not restricted to any certain type of matter. All masses, hot, cold, metallic, or non-metallic are affected by it. As Galileo demonstrated in the 17th Century, under the effect of gravity, a feather, in a vacuum, will fall to earth at just the same rate as does a cannon ball. All materials, of whatever kind, large or small attract one another to the same degree; proportionally as to their masses. A drop of water, falling free in space, immediately forms itself into a sphere, and will attract any other nearby drop and merge itself with it into an even larger sphere. But, how does it do this? How does it attract the other water drop? How does it even "know" the other drop is nearby?

The answer to this question is, again, to recognize the universal powers of the electrons to act "At-a-Distance." Gravity exists because the electrons of any body, great or small, are always attempting to re-establish what they feel should be their "correct" distance between one another. Just as in the transmission of electricity along a conductor, when electrons sense they are "too close" they move apart to re-establish their correct intervals. So too, in the macro-universe, do the electrons of all forms of matter sense, over any distance, that there are other electrons, in other bodies, at a greater distance from themselves than what is the optimum. Thus, they mutually attract one another in an attempt to merge into a new spherical mass at their optimal intervals. We see, therefore that gravity, like electromagnetism, radio or light is not material but is, rather, the transmission of "knowledge" from one set of electrons to another. The electrons of both the Earth and the Moon "know" that the other is nearby and, if they could, would combine to merge into one single planet. The Moon is continually falling toward the Earth and the Earth is continually falling toward the Moon. The only thing preventing them from combining is the Moon's relative velocity and its circular orbit that provides the centrifugal force necessary to keep them at their present distance from one another.

Astronomers say the Universe is expanding. This is reassuring, for if it were not, all the electrons in the Universe, by attracting each other, "Acting-at-a Distance," would, finally, bring themselves back together into one single cosmic mass; creating, perhaps, another "Big Bang."

ORGANISM

Electrons within organic molecules impart electromagnetic charges to all living cells. The processes of life such as osmosis, sight, reproductive, muscular and nervous activities, all involve the transfer of electrons across a potential gradient. Each cell has an electrical charge, either positive or negative; a quality utilized in the laboratory process called electrophoresis; by which cells in solution may be separated into positive or negative types by passing them over slanted metal plates charged with a weak, direct, electrical current. This test is also used in the testing and analysis of serum DNA. The charge in the cells is maintained by a battery-like effect between the salt and potassium ions, in solution, within all cells. Collectively, these charges combine, positive to negative, to maintain a strong electrical potential within the body that might be considered as part of the permanent "life force." Our living bodies are, literally, bound together within a powerful magnetic field by the cells' electrical attractions; by the mutual attraction of the positively and negatively charged cells. These opposing polarities bind our bodies within an intense electrical field that is dissipated only by death.

When studying the development of embryonic life, biologists, often, observe the growth of chicken embryos through transparent windows placed into the shells of fertilized eggs. Within hours after fertilization, what they observe is not replications of the original cell but the spectacle of thousands of now specialized cells racing in every direction along paths that, eventually, will become the arteries and veins of the future chicken. The cells work feverishly, according to some "known" plan and purpose. What can the purpose be that these cells "know?" The thing they "know" is that their mission is to create a chicken; and a chicken it must be or they must die in the attempt. The accepted explanation for this knowledge is that the plan of each life form is encoded within the original cell's DNA that, then, automatically reproduces and differentiates itself thereafter. But, while the DNA may be the matrix, it does not explain life. A cell with its chemical DNA intact would still remain an inanimate chemical object without the "drive" and the "knowledge" possessed by each cell within a living organism. From whence comes this "knowledge" and "drive"?" Certainly, in the case of a chicken embryo, it cannot have come from a brain that has not yet been developed. But we must ask: If this "knowledge" has not yet been developed internally, from whence does it originate? If not internally, this "knowledge" must have been implanted from outside the organism at conception. Could it be that each embryo, whether plant animal or human, is implanted with its own "soul" by means of "action at-a-distance; each given its own electromagnetic life force as its program and guide for the remainder of its life? Is this unbelievable? Perhaps it is. But this is the direction the evidence leads. Let skeptics disagree, but, we ask them to provide a better answer.

Why is the possibility of "Action at a Distance" dismissed as fanciful by professors who write articles to oppose lines of inquiry they describe as "scientific fallacies?" Is it done for some higher purpose? The idea that the electrons seem to have the ability to sense information, At-a-Distance, is opposed precisely because this line of reasoning opens up a "Pandora's Box" of ideas leading toward "superstitions,"—superstitions such as the existence of God as the "Force" underlying the basis of all living matter! Once accepted, "Action at a Distance" might even explain how God communicates with Man and how Man communicates with God. It might "scientifically," explain the power of prayer.

Is this not what these pseudo-scientists fear and why they turn away their faces? Is this not why Professors Piccione, Mehlhop and Wright, state in their letter to the editors of <u>Scientific American Magazine:</u> "that the acceptance of the idea of "action at a distance" would set science, (in their view), back for hundreds of years." They sense only too well what those ideas infer. Are there classes of evidence scientists must not consider?

To counter these thoughts other scientists now try to explain the "Action-at-a-Distance" phenomena while, maintaining a mechanical viewpoint, They propose a new solution called "String Theory." This posits that, throughout the universe, there may be innumerable strings of physical connection, that perhaps exist in another dimension and that these "strings" are what, physically; connect the electrons of the planets and all other matter. This theory attempts to explain "Action-at-a-Distance; in physical terms. Unfortunately, these "strings" have yet to be proven to exist.

A deeper understanding of reality should take into account the "knowledge" of the electrons. Electrons *do* "sense" each other and they *do* exchange "information;" as in the examples of light, radio and gravity, by "Acting at a Distance." Accepting the idea that electrons interrelate with one another, At-a-Distance, does not signal "the end of science as we know it;" quite the reverse. It means the extension of human knowledge; a broader view of reality. Responsible science is coming to realize that a science based only upon mechanism is bound by a self-imposed limitation.

Unbiased investigators are led toward a realization that the universe is material only in gross. At the atomic level and beyond, what is found is not material; but electromagnetic charges; light, gravity and perhaps other, yet to be discovered, forces. Accepting "Action at a Distance"—and seeking to understand how it operates, will dramatically advance the frontiers of knowledge. As in the time of Galileo, disturbing discoveries may be suppressed for a time as politically inconvenient but, eventually, "Action at a Distance" will not be denied. Its existence is suspected in too many quarters. Truly, this thought *is* on the wind. Acceptance of this concept will not imply a retreat from rationality, nor will "our science" be set back for hundreds of years." On the contrary, it will release us from the shackles of materialism and open up new vistas for scientific inquiry.

© Les Crane 2007

"CULTURE & BELIEF"

A CHRISTIAN OVERVIEW
OF HISTORY

©LESLIE C. CRANE
2006

CONTENTS

PREFACE

The former chairman of the American Airlines Company once said that; "from a height of 30,000 feet, human beings and their works look just like ants." This observation, if taken, as literally true, might discourage one about the status of humanity. But could it be so? Are we little more than the ants who believe themselves supremely important as they wage their meaningless struggles over who will control the world's last ant hill? This is a depressing thought. But upon hearing from afar the notes of a Beethoven Concerto or a vocal duet from a Verdi opera, we are reassured. The ants, we reason, could not have created those works of surpassing beauty. This may be the proof we need to reassure ourselves that: no, we are not ants; we are higher beings; each with an individual soul which, however dimly, senses there is a purpose to life. History is the record of our instinctive upward-reaching toward discovering that purpose.

Mark Twain said that history is just "one d—d thing after another." Henry Ford added that, "History is Bunk." Neither of these opinions is true because we sense an underlying force behind the flow of historical events, not due to chance. Past and present historical developments were and still are the result of planning and direction. Our goal is to describe those factors so that the meaning of the past may be correctly apprehended and become useful in improving upon our futures. In making this effort, it is necessary to name the persons, forces and organizations involved.

Regarding the evils of this world, who of us is entirely innocent? We all have erred and been deceived. Frankness will be welcomed by the common reader who seeks for meaning behind the confusion of events. They want plain truth. Others, the pseudo-elites; those who use their ill-gained positions to rule over their fellow men, essentially, by cheating at the game, they will be confounded. To them we make no appeal.

Christians, Protestants, Catholics, Muslims, Jews, Agnostics and whomever; should approach these topics with open minds, for all are vitally affected. None are immune to being deceived. In the following chapters, we present our evidence as the facts do lead. We neither praise nor condemn individuals. We seek to comprehend the underlying truth and to, thereby, provide ourselves with a more complete understanding of our past and of the situation in which we find ourselves today.

All of us are targets in a deadly psychological war; a war being waged against us by an unseen hand; an enemy that forces delusions upon us; delusions designed to bewitch and to enslave. This is the Force from the "dark side." Continually, it employs the technique of "divide and conquer." It sets nation against nation, race against race, religion against religion, the young against the old, men against women, children against their parents; its crimes are endless.

But, while describing these deceptions, we offer, also, a solution. In this we follow in the footsteps of that great Puritan poet, John Milton.

Sonnet on His Blindness
1655

Cyriack this three years day these eys though clear
To outward view of blemish or of spot;
Bereft of light thir seeing have forgot
Nor to thir idle orbs doth sight appear
Of Sun or Moon or Starre throughout the year
Or man or woman. Yet I argue not
Against heavns hand or will not bate a jot
Of heart or hope; but still bear up and steer
Right onward. What supports me dost thou ask?
The conscience Friend to have lost them overply'd
In libertyes defence my noble task
Of which all Europe talks from side to side.
This thought might lead me through the world's vain masque
Content though blind had I no better guide.

CHAPTER ONE

THE POINT OF DIVISION: EVOLUTION

"In the Beginning God...."

History began with the Creation. God's Creation is fundamental to all that follows. God's enemies recognize this fact and make His Creation their central point of attack. The controversy between the biblical account of the Creation and the Evolutionary doctrine must be resolved before proceeding.

"In the beginning God created the heavens and the earth." (Genesis I, 1) This is the beginning of history. It is all we know, all we need to know and finally, in our present state of mental development, all that we can know of the Creation. The vain imaginings of those who pretend to be "scientists" but who in fact consider themselves to be "demigods" fail and always will fail to expand upon that simple account. Every generation sees new theories advanced that pretend to explain the origins of life in rational terms. Ever-new scenarios are fabricated and for a time gain currency with simple minds. Yet, while each succeeding theory enjoys its moment of fame, soon it is run off the stage by the next professor's even more elaborate fantasy.

Any honest pursuit of science should be the exploration and the unfolding of God's wonderful universe; followed by the technical application of the knowledge gained, thereby. In a broad sense today's applied science does this. It provides a wealth of applications that create higher and higher technologies. But with regard to the Creation, a pseudo-science has grown up; driven by a religious motivation that only calls itself "scientific." Evolution is in fact no more than a bundle of assertions advanced to counter the Word of God.

To spearhead their attack, a Theory of Evolution was invented by these followers after Satan. They said that; "of course God did not *really* create the world in six days." Questioning God, they imagined that the process must have begun without any causation at all—accidentally. This is the proposition that life created itself! To them it followed that if there was no special creation, ergo, there is no God. That was their true goal. It stated that living cells must have created themselves by their own volition and thereafter "evolved" into all the forms we observe today. With no proof whatever of this process, evolutionary "scientists" cover the weakness of their claims with another lie; the lie that the creative process took "billions and billions" of years so that, while they are all-knowing, they cannot be expected to provide any detailed supporting evidence at this late date—a most convenient evasion.

Evolution is asserted as the only "logical" explanation for the existence of life; which it might appear to be to those of limited understanding. Therefore, no further proof is necessary. Evolution must be accepted *without proof.* To be "politically correct" modern scientists, when confronted with such an unproven proposition, find themselves forced to make a "leap of faith"—that "leap" required of them to believe in Evolution; proving that Evolution is a religious dogma and not a proven fact.

Whether one believes organic life came into being by accident and then evolved into the life we see about us today or whether one believes that the world and its life forms were created by some unknown mysterious external agency; is the central question facing the world today. Philosophically, it *is* the "point of division." The debate over Special Creation vs. Evolution defines the two, irreconcilable, opposing world-views. There is no middle ground. One's position on Evolution determines one's thinking on every other scientific, social or political question.

THE DOCTRINE OF EVOLUTION

Lie # 1-God did not create all living things.

Explaining the origin of the Earth and its inhabitants, the Bible states: God said: "Let the Earth bring forth grass the herb yielding seed *after his kind and* after creating the sun, moon and the stars God created great whales and every living creature that moveth which the waters brought forth abundantly *after their kind* and every winged fowl *after his kind* and God saw that it was good. And God made the beast of the earth *after his kind* and every thing that creepeth upon the earth *after his kind*: and God saw that it was good. So God *created man in His own image (after his kind)* "In the image of God created He him. Male and female created He them."

Lie # 2-God did not create the separate species and types.

But obviously God, or whomever, *did* create every plant and animal in their individual species. We cannot know the date or time for that is not given and speculation is vain; but proof of the creation of *separate* species each *after their own kind* is obvious. At this time a horse and a cow cannot be crossbred and the argument that sometime in the past their predecessors could have been has now been refuted by DNA analysis. Their differing DNA conclusively proves that the horse and the cow *never* had a common ancestor. Furthermore there are not to be found any of the intermediate fossils in the geological record between early and present day species necessary to prove the ongoing process of evolution. (Where is the horse-cow that should be able to breed with both true horses and true cows?) The evolutionary argument fails, absolutely, because there are no intermediate types between the species. If evolution were true there would be no separate species, at all, but only one; undifferentiated mass of creatures and plants with but gradual differences between the individuals; each being able to breed with its, adjacent neighbors; above and below it in the chain of life. The often used example that purports to show the evolution of horses from breeds which were small into breeds which are large is not a proof of evolution—for, genetically, all were horses varying only in size. The Creation of the ordered universe and of the organic life we see all about us is the result of the act of some intelligent force or supernatural being totally beyond our comprehension; best personalized by the word God. This is as obvious and as proven as any fact can be, for us, at this time.

Nevertheless, God's enemies rest not. Their lying assertions continue step-by-step and point-by-point. Today, to be published, the authors of any scientific book article or television script are obliged to begin with the statement that—"In the evolutionary history of the *XXXXXXXX* we know that....!" They make this obeisance to the icon of evolution because, well they know that, without it, their grants and sinecures will evaporate—that they will be silenced, unpublished and quite possibly unemployed. Today's "new-science" is totally thought-controlled.

Lie # 3—God did not create men and women in his own image.

The underlying purpose of this assertion is once again, to deny the word of God and to reduce the status of human beings into that of beasts; not the special creation that *God created in His own image.*

By the logic of this lie, the human race has no special rights or mission in the world and it might, in fact, be defined as parasitic. The highest forms are simply those dominant in the struggle for survival at any given moment. The dominant life forms, eventually, may come to be the ants or the rats. It makes little difference to an evolutionist. His motivation is the hatred of God. His ultimate goal is the negation of life because it *is* of God's creation.

The propositions contained in the Theory of Evolution are utterly unproven but yet are blindly accepted by fools and dogmatic atheists; while the opposing theory, Special Creation, proven by true science and by common observation is denied out of hand. How the Creation was accomplished is a question that cannot be answered by the powers of the human mind. None knows the mind of God or understands His purposes; but we may study and enjoy His universe; for, "the very heavens *do* declare His handiwork." True scientists are not those who war with God but those who cooperate with God and honestly seek to explore and to understand God's Creation.

Evolution was purely a religious dogma put forth by the enemies of God during the middle of the 19th century. It has been continued as a theory on through the 20th and into the 21st century primarily to support a godless social agenda. These plotters mask themselves under many names such as "Fabians," "Enlightened Ones," "Deists," "Humanists," "Socialists," "Fascists," "Communists," "Progressives," or now as "Liberals and Democrats." In our own day, they continue to advance their agenda; the planned subversion of all the existing world cultures and of all divinely-given moral laws. Their leaders have no personal belief in the tenets of Evolution. To them, it is but a weapon, a weapon useful in achieving their higher goal—the founding of a worldwide godless political state; under conditions of social totalitarianism; over which they intend to rule. They are in fact planning a New One World *Dis-Order*. A *Dis-Order* in which they imagine they will be the masters; ruling, not with the counsel of God, but with the guidance of *their master*, Satan himself.

Have we not eyes to see? Cast them about for proof. The root cause of all the wars and atrocities imposed upon the innocents of the world over the last two thousand years as well as the suffering of the hundreds of victims being martyred now, each day, can all be traced directly back to this, God hating, intellectual movement.

In the 1930s, the American folk-singer, Woodie Guthrie, sang a siren-song which said; "The answer is on the wind" Yes, that answer, godless Socialism, *was* on the wind; and the world now groans in agony as it reaps its whirlwind.

CHAPTER TWO

CULTURAL HISTORY

True history is Cultural History. It is "The History of Ideas."

History should not be understood as simply a narrative of events. Meaningful history includes a deeper exposition of the then-existing cultural conditions along with the motives and purposes of the participants that, at that time and in that place, determined their actions. Historical events are dependent upon beliefs and motives. These were determined by the intellectual and cultural forces at work in a given time and place. We cannot comprehend a historical cataclysm such as the French Revolution by simply memorizing a list of important dates; neither by analyzing the acts of the Convention nor the deeds of a Napoleon Bonaparte. While the facts may form an outline, the story of the French Revolution lies deeper. It is found in an understanding of the ideas operational in the minds of the participants, in that time and in that place. Knowing these, we find the acts of the participants came forth into actuality with the force of destiny. This is the knowledge we seek; an understanding of motives and ideas. Only these will satisfy our longing for completeness.

This cultural approach to history was outlined and developed into the fullness of a grand historical system by the one who must be acknowledged as the most profound cultural historian of ours or of any other time; the German philosopher Oswald Spengler.

His masterwork: "The Decline of The West;" ("Der Untergang des Abendlandes"), stands as the intellectual monument by which all further histories must evermore be measured. Conceived during the beginning years of the 20th Century, it did not address events of that period but, rather, it had a timeless purpose. Published, worldwide, in the 1920s "The Decline of the West" made a profound impression upon the intellectuals of that day. Historians were astounded by the scope of its thesis; and although they made some attempts to criticize the work in its minor details they were, soon, forced to acknowledge they had been overshadowed by the work of a Titan. Its theme is so convincing it forms the mental framework of all later seekers desiring the deepest understanding of history. Spengler provided an Olympian overview, of the world's cultural history from a height never before surmounted.

He argues that human cultures develop in an "organic" manner; as do living organisms; having, at their beginnings, a set of unrealized potentials similar to the genetic code of living cells. Thereafter, these seed potentials are "actualized" through time into becoming the artifact we define as a culture. This idea once apprehended becomes undeniable.

Using as examples, civilizations for which we have good historical data, the author describes the process by which each separate culture has, in their development, passed through identical phases. Defying the arbitrary classification of world history into the divisions of "Ancient" "Medieval" and "Modern," the author sets up a new standard; unrelated to a single time-line.

Each culture has had its own time-line without regard to the time-lines of others. Cultures may indeed coexist but yet, individually, be at different stages of their life spans. In 1000 BC when Egyptian culture was beginning its decline Graeco-Roman culture was only at its beginning. Spengler demonstrates exact parallels between the developments in several historic cultures. The Classical (Graeco-Roman), The Magian (Judeo-Christian-Arabic), and the Faustian (Gothic Western-European-American) showing that each of these cultures has passed through, as ours is passing through today, identical political, intellectual and artistic stages.

Cultures in their Spring-Times are filled with unrealized longings; with an understanding of hidden potentials. Those potentials become actualized over time into a finished cultural system of living. As each culture completes itself, it finalizes into its fully realized and frozen state of "civilization," the state of intellectual rigidity and death. This is the period wherein a culture's flowering is complete; its genius spent; it is the period during which it may continue on as a living form but wherein cultural and technical expression becomes merely replication and extension; without further advancing its fundamental thought or goals. Completed, it may be supplanted by a younger culture or live on in lingering death in a "fellahin" state best exemplified by the present remains of the Egyptian Empire which, for the last two thousand years, has been overshadowed first by Magian and then by the Faustian culture. During this period of lingering death there may still be repetitions of the old forms but further cultural advancement is impossible.

Chinese civilization lived on for centuries in an extended post-classical period. Having fulfilled its potentials, Chinese culture "died" and became a cultural "corpse" fixed in an endless repetition of its social and artistic motifs until it became prey to a younger culture. That event occurred when the Western "Faustian" culture arrived upon its borders, in force, during the 19th Century. Now as we see, a new culture is developing as the Westernization of China proceeds at break-neck speed.

Japanese culture provides another illustration. It too was a completed system fixed in the sleep of feudalism until "awakened" by its collision in 1847 with the American, Admiral Perry, and his fleet of warships. Japan, likewise, has been partially Westernized and it is now developing a new Japanese-American cultural form.

Spengler illustrates his analysis by comparing the development of historical cultures with the lives of plants, animals and human beings. Deriving his ideas from the philosophy of Goethe, he believes in the idea of Teleology; of the Destiny of Organisms. He asserts that cultures, just as living forms, have a "purpose" contained within themselves at their very beginnings.

Just as a plant's seed has within itself a purpose, which is to grow, to flower, and when its mission is complete, to form its seed and then to die. So it is with a human child who *must*, unless it sooner dies, grow into a man or a woman who likewise will reproduce age and finally die. All living forms pass this way. A culture's period of growth is defined by Spengler as the "Becoming" and after fulfilling its destiny, as the "Become." In art, a work during its creation is a thing "Becoming;" but, when finished it is a thing "Become." Human Cultures are living organisms composed of a certain people in a specific place and time. Their inherent potentialities are self-contained at their seed-time and grow outward through youth, into maturity; through self-realization into fulfillment. When the process is accomplished the culture is complete—"the Becoming" is now "the Become."

Parallel developments occurring within different cultures may be compared. These parallels are described with illustrations drawn not only from politics but also from the arts of music painting sculpture and mathematics. The time periods of those cultures which existed or still exist in the Mediterranean-European-North American regions of the Earth are approximately as follows: The Graeco-Roman Classical from 1000 B.C. to the year 1 AD, the Magian Period from the year 1 AD to 900-1000 AD, and the Faustian (Gothic-Western) from 900 AD to the present. As these societies grew to self-realization, their political and artistic flowering followed identical paths. Their arts politics, science, and mathematics were not identical in content; as each Culture was driven by separate underlying concepts but the sequence of these developments remained the same.

The flowering of Classical and of Faustian mathematics and their associated musical creations occurred at the same stage in both cultures. The Greek conception of mathematics was based upon a perception that required the counting and relationships of "real" objects. Numbers for them described objects; the lines and planes of Euclidean geometry defined real surfaces. Of the mathematics of imaginary functions they knew nothing and desired not to know. In fact their natures recoiled from such concepts.

Faustian mathematics in contrast is concerned not with the numbering of "things" but rather with the study of relationships such as those described by trigonometric functions whose most characteristic concept is an abstraction—a moving point in space. Their separate, instinctive, feelings produces two separate mathematical and artistic systems; but at the same stages of their development.

Politically and socially, the Greeks insisted upon a rigid and unchanging system; whereas Western culture is comfortable with a flexible order subject to continuous amendment. The Greek attitude is well illustrated by the theme of the play "Medea" by Euripides. Jason the king of Athens returns from his quest for the Golden Fleece a hero but brings with him a foreign wife, Medea, to reign with him as his Queen. Under the moral order of Greece marriage to a foreigner was forbidden because it challenged the laws of the gods. This violation led to the destruction of Jason and of his Greek wife and children.

Conflict between the errors of men, versus the verities of tradition, is the moral teaching contained in Greek tragedy. Those plays concern themselves with the defects of human nature as expressed in the emotional excesses of the principal character(s). Those emotions lead him or her to violate the moral order while, in the background, the chorus (the polis) chants their warnings—witnessing to the crime; forecasting doom. At the end, the gods must intervene to destroy the miscreant and to reaffirm the traditional order. The Greeks believed they had to live by the established and immutable laws as given to them by the gods; while Western parliaments enact new laws each year as conditions dictate

Similar contrasts may be seen in architecture. The classical form demanded that the plan be heavy; wider than its height; balanced, symmetrical; rooted to the earth. However, in Faustian culture, during the years 1000 to 1300 AD, our so-called primitive forbears began to pile up those great Gothic cathedrals in Western Europe whose beauty we marvel at yet today. These were built higher than they were wide—with each succeeding attempt reaching ever higher 'til stone could bear no more. In one final overreaching, in their highest example, the cathedral at Beauvais, collapsed into what remains today a noble ruin. But as we see, the remaining cathedrals still stand; they still reach toward the heavens. Had their builders been able, they would have floated them on air—such was their wish to conquer space in their upward striving towards God.

That same desire lives on in our modern Faustian souls. Because Earth has substantially been explored, our dream is to seek out even greater heights. Our quest, we think, must be the exploration of interstellar space; exploration to be endlessly carried on by our Faustian astronauts in the future.

Such longings and felt purposes were pregnant at the foundation of our culture. They were within our seed potential and formed our culture over time. For us the last 1000 years has been a period of actualization. But now, as the possibilities of our Gothic potentialities essentially have been worked out by every means possible; in cathedral building, in the visual arts, in music, in science and in political systems; all that was originally possible for us to imagine has been actualized.

We must accept that those goals were peculiar to our culture and were not the goals and desires of some other culture. Spengler's title: "The "Decline of The West" in reference to our present culture is not based upon unfounded pessimism but upon an honest appraisal of the cultural-historic stage at which we find ourselves today. Spengler's is the cold-eyed recognition that our culture has come to its late and final period. Faustian arts and imaginations have matured into full flower. Completed, its development can go no further. Beyond this time, for us, there can be only a hardening into the state of the Become. This stage of a culture he calls Civilization; it is a period, technically glorious, but consisting essentially of repetition and extension. Politically, its form is becoming Caesarism.

According to Spengler, no longer can there be any hope of a further spiritual upward development. This is illustrated by considering the present state of our arts of music, mathematics, architecture and literature. The potentialities of Western music were fully realized by the late 18th century in the works of Bach Handel and Mozart and Beethoven. Since then, while we have seen periods of romantic elaboration and repetition within that system we have not seen any fundamental, further, musical advancement. The works of Liszt, Chopin, Schubert, Verdi, Rossini, Tchaikovsky and finally Wagner, Brahms, Richard Strauss, Ravel, and Debussy while all must be granted the highest of praises; all are derivative, based upon an already finalized musical system; now carried on as the "classical tradition."

By the Twentieth Century the fires of that "tradition" had burned low and only a few new examples were to be created. The works of the late Russian composers Stravinsky, Shastokovich, and Prokoviev remain Nineteenth Century because, as we know, Russian culture lagged the Western by about a century and was artificially frozen in time by political decree. Now, opened to Western influences the time for new Russian symphonies is at an end.

"Classical" orchestras now devote themselves to the performance of works written from 100 to 400 years ago. These constitute the only body of music understandable to our ears. When attempts are made to write "modern" music or works based upon new tonal systems, the resistance of audiences is adamant. Such works to us are "noise" and un-listenable. Every effort to expand our musical envelope becomes more and more feeble. Would-be composers and symphony conductors sense this lack of advancement and complain that new classical music is not being written; but find themselves unable to create it.

In mathematics as in music the final development of our concepts was realized in the 18th Century by Descartes with his discovery of the calculus, by the discovery by John Napier of logarithms and by the possibilities of multidimensional geometries as discovered by Gauss in the early 19th century.

Today, while our facility in handling mathematical calculations is made instantaneous by the electronic computer, the principles upon which we compute remains those of the 18th and 19th centuries. It would appear that, with the set of *our* minds, we cannot improve upon that mathematic and feel no need to do so.

With the visual arts we see the winding down of the commitment that found its climax in the art of the Baroque. By the beginning of the 20th century while artists had the same facility as those of the past, they had lost their purpose and direction. Today while a great deal of artistic illustration and "busywork" continues to be produced, the artists of our time have little to say. Frustrated, many turn toward abstraction looking for meaning in geometrical forms or in the contemplation of their own psyches. In such a period of decline, the works that command a money premium are often those created by frauds, mental defectives or alcoholics who drip house paints upon canvas' laid flat upon their studio floors. Self-promoting drug addicts now create photographic reproductions of soup cans, cartoons and other common objects. For this they are lauded to the skies.

As their audiences are equally uncritical, these "artists" find it profitable to become the charlatans of art in pursuit of money. The famed artist Pablo Picasso, who was fully capable of producing honest art, confessed, late in his life, that the times no longer called for traditional painting. Those forms had been fulfilled. The old ways had died. Therefore, as the market wanted abstractions inanities and political symbols well then he would give it abstractions' inanities and political symbols; charging all that the traffic would bear. Those who still seriously search for form almost universally look to the past. Their landscapes must contain decayed 18th or 19th Century ruins; relics of a better time. Their human figures are obliged to be romanticized Polynesian, Amerinds, or other primitive peoples who satisfy the romantic fallacy that modern man is debased and therefore primitive peoples must inherently have been more noble than we; symbols of a purer past. Such is the looking back; the longing for a lost time when; as the artist imagines his work and his spirit could have been as one.

There is a profound loss of purpose in the minds of certain modern artists who sense the angst of our time. An artist who paints 50 square feet of canvas as a blank sheet of color depicting absolutely nothing *is* honestly painting that which, for him, is the truth; the truth of absolute skepticism. He believes nothing, he feels nothing, and therefore he paints nothing; in a futile act of non-creation.

Typifying the nihilism of our times is a work by the Norwegian artist, Edvard Munch, entitled "The Scream." This famous painting presents a desolate landscape with a distorted face in the foreground, its mouth agape, emitting a primal scream; a scream of utter horror. What can the meaning of this painting be? What is that horror? Is it not the realization that the subject and the artist himself, having searched high and low through the philosophies of our time, have finally found nothing—nothing—that cannot be questioned. The artist and his subject find, at the ultimate juncture, no human value worthy of belief; in all this world—nothing. With one final act of skepticism they peer into the mouth of Hell and start back in horror. What else can this painting signify?

This mentally disturbed artist's works are now so cherished, by the morally bankrupt connoisseurs of the present art world, that a special museum has been built to house and to protect them. Paintings such as these are the symbolic final acts of a spent tradition; a tradition with nothing left to say.

Literature follows a similar path. First comes a period of slow internal growth during which the possibilities and purposes are considered leading; on to a flowering that contains as its motifs the ideals of style founded upon accepted principals of intellectual restraint and moral justice. Later, come periods of questioning and experimentation. At the end comes skepticism; which leads to a denial of all former principals and beliefs.

Onward and downward lead the topics of our popular writers. Rather than looking outward into the world around them, they become introspective; searching for a new order they imagine can be found in the study of human mental problems.

The late 19th Century witnessed the development of the "problem novel" and "problem play" that presented real or invented "problems" that the authors asserted afflict the human condition. Ibsen's play "Hedda Gabler" is concerned with the "liberated woman problem." Like today's dramatists, Ibsen presents his "problems" without answers. In the final act of this play the author concludes that the problems of Hedda Gabler are insoluble, so that her only option was suicide.

Today, all sense of purpose has been lost. Authors with nothing left to say deliver "stream of consciousness" accounts of their meaningless daily lives. When these fail to entertain they lower themselves further into writing vulgar-minded pornographic plays and novellas; finding in these topics the only remaining basis upon which they and their demoralized audiences can find communion.

Religion Determines Culture. Societies based upon a religious code of morality produce relative peace and social justice. Societies based upon the slippery ideas of human logic, in contrast, slide inevitably downward into the slough of Socialist Imperialism. That is the final stage of *our* culture. That is where we stand today; in the state of the "Become;"—"Hyper-Civilization." Truly, this *is* the "Going Down of the West;" paralleling exactly the decline and fall of the first Roman Empire.

Time scale: Ancient Rome 100 AD.= Faustian 2000 AD.

CHAPTER THREE

FIRST CULTURES

And God said "Let us make man in our image after our likeness: and let them have dominion over the fish of the sea and over the fowl of the air and over the cattle and over all the earth and over every creeping thing that creepeth upon the earth." "So God created man in his *own* image in the image of God created he him male and female created he them." (Genesis I. 26, 27)

Note: that while God created man and woman in his own image he did not create them as lesser gods capable of self-perfection into the status of co-equal gods. (A pernicious notion which is the basis of many "pseudo religions and rites of perfection") Men and women are less than God; created not creators; less than God because they are capable of choosing evil as well as good. This is the meaning of the words "original sin." Original sin is not caused by the commission of a specific sinful act but a *condition* of moral imperfection; as having the ability to sin.

After the Fall, when the Angel drove the first family out of Eden they and their descendants, with no previous models to follow but being as intelligent or as ignorant as we would have been under the same circumstances, scraped by as best they could. They hunted for food gathered fruits and finally learned the skills of settled agriculture. As seemed natural to them they grouped themselves first into families then tribes and as collections of tribes into nations. When their agricultural effort produced a surplus they built towns and cities.

The advance from primitivism to city dwelling occurred not over a period of millions of years but, immediately, and within a few hundred years. It is a fundamental fact of prehistory and remains so in the present that a people whose economy is based upon hunting and gathering will never build permanent cities. But, when and wherever, an agricultural economy is established the founding of towns and cities begins immediately. These two economic systems need not develop sequentially but may co-exist in the same time period. Because the hunter-gatherer way of life produces no surpluses its members are condemned eternally to pursue their sources of subsistence as the nomads of Central Asia do today.

In America, while the, agriculturally based, mound building Indians of the Eastern United States were living in permanent villages and towns, their contemporaries, the Plains Indians, never considered setting one stone upon another.

City building is inevitable as soon as agriculture becomes established. Agriculture produces a surplus and the city becomes its market. Agriculture is the root, the city the flower of any established culture.

Wherever we discover the remains of an ancient civilization we turn up not the huts of farmer-peasants but rather the ruins of highly sophisticated stone-built centrally planned cities. Whether it be in the ruins of Mohenjo-Daro, in Pakistan, Teotihaucan, in Mexico, Palenque, in Guatemala, or at Uruk on the Euphrates; we find the same identical plan employed—straight streets laid out in a right-angled grid. The straight line, the right angle and their resulting patterns are as instinctively necessary to mankind as is the hexagonal plan to bees and wasps. The straight line and the rectangle are, in a very real sense, the mark and motif of the human species. Our forefathers were no less intelligent than ourselves and were motivated by the same inwardly felt necessities as we today. The design of their cities confirms that, in essence, New York and Babylon are one.

The first cities were founded along the river networks. They soon became centers of power; drawing into themselves the wealth of their surrounding territories. Next, there was a felt need for a centralized system of government. As seems inevitable in human society this resulted in a system of authority requiring a single head, a king or dictator holding power by elective or hereditary right. This figure reigned by his physical strength. But resistance to his authority frequently developed; both from the people and from the envious among the aristocracy. It soon became apparent that men could not be ruled by force alone but that kingship needed some added authority. This it was found could be acquired if the people could be persuaded to "believe" in the "divinity" of those self-appointed kings. Linked with a priesthood, the king's power was made absolute by the controlling of men's minds employing the power of religion.

Men and women naturally seek to know God. God's Word is written on their hearts. Instinctively they know there is a difference between good and evil, right and wrong. Whatever their condition, men feel the presence of God and desire to follow his will. This human longing for God's will has been exploited by the governing powers that, ever and always, form themselves into the priesthoods of the false religions. They place themselves between God and man. By interpreting God's will; they deny men their rightful access to God. Total government is psychological government; the fusion of Church and State.

Meaningful history must take into account this alliance of Church and State recognizing that behind every major historical development, past and present, is the religious motivation. Religion determines culture; not the other way 'round. Spengler's failure to include the dynamic of religion in his comparison of cultures left his otherwise remarkable system of analysis without its capstone. Unable, or perhaps fearful of the then current intellectual climate, he failed to assert the possibility that religion, too, was a force in determining culture. Rather, he substituted a "racial determinant;" that a race of people, with a "given" set of ideas, would cultivate these fundamental ideas until they were fully realized. Those initial ideas would control the culture's development; from its seed-time, through its flowering, and on toward its fulfillment.

Spengler describes human cultures as living organisms; directed by their "destinies." The seed *must* become a plant; a "people" *must* become a culture. Their destinies develop during the process of "Becoming" until they are fully realized in the "Become." A "people" in a specific location having a high potential within itself has a sense of purpose that commits it to grow and develop in a certain direction.

In Spengler's view, within the peoples of early Greece who had been placed by fate about the Aegean Sea there was the felt desire to one day build the Athenian Acropolis. At the same time in another place, in the foggy forests of Germany, the members of another pregnant culture felt within themselves, as the future Franks and Germans, the inward necessity to one day build the Gothic cathedrals.

But is this entirely so? Or might a culture's destiny be modified by its religious beliefs. Should it not also be recognized that the designs of the Greeks' temples were dictated not only by their racial drive but that these structures were, in fact, symbolic representations of their religious philosophy? The phallic columns depict the physical passions of the fertility cults whose rites were anciently celebrated in the "sacred groves." But added to this obvious element, the columns were then capped o'er by the powers of geometry, representing reason, in the form of the triangular tympanum. These motifs; column and tympanum embrace the two contending elements in Greek religious thinking. Was this development inevitable from the beginning of Greek history? Obviously it was possible; but it was not inevitable without the combining of their dualistic religious motivations.

Europe's Gothic cathedrals and Handel's "Messiah" are also symbolic works expressing the power and grace of the Christian faith; but this was a faith unknown to the proto-Gothic peoples. Faustian culture flowered as it did not only because of its inner necessity to develop racially in exactly such and such a manner but, rather, it flowered following the merger of the destiny of a strong and determined people with the additional dynamic of the Christian Ideal. It was the synthesis of a "people" in one location *with a new religion* that created a new Gothic destiny; an outcome neither predetermined; nor inevitable.

In the past, the intertwining of governments with the established religions has meant that much of the surplus-wealth of these cultures was diverted into the creation of monumental temples and public works. Within the designs and decorations on these monuments are to be found symbolic clues that supply the keys to an understanding of these cultures' religio-political systems. We deeply sense there are hidden meanings locked up in the art and forms of ancient monuments such as the Egyptian Sphinx and the Pyramids; although their full significance remains unknown to us.

ATLANTIS

Archaeology deals predominantly with the classification and study of objects found buried in the earth because these are what has survived the damp and rot. Therefore, our thinking about past ages comes to be based upon these stone, metal or pottery remains. Yet, as archaeologists sift layer by layer through the soil attempting to research an ancient culture by collecting bones and pot-shards; and with their noses pointed toward the earth; beside them stand massive temple ruins which if even half-deciphered, would reveal far more concerning the culture's ideals and motivations than the collecting of the micro-remains of common everyday living. Nowhere, has this been more true than in attempting to understand the motivating factors of what, erroneously, is called the Pre-Columbian culture of Central and South America. This culture more properly should be named the Atlantean culture.

Only gradually over the last 150 years have archaeologists come to an appreciation of the size of this culture. Its monuments are spread over a region extending from Central Mexico downward along the South American Pacific Coast for a distance of 2500 miles. Here lie the remains of vast cities filled with massive buildings far surpassing in quality and extent any structures known in Europe or the Middle-East before the rise of Egyptian engineering. These buildings are the work of populations that must have numbered in the millions. The density of the populations surrounding the cities must have been thousands per square mile indicating a division of labor that specialized in many occupations including intensive agriculture and trade. Today's airline pilots flying over the jungles of Northern Columbia at sunset or at dawn see in the horizontal shadows evidence that large areas, now only jungle, that were once cultivated with terraced canal and island-block systems of farming. The remains of these canals and enclosed fields are still evident. Indeed, this method of farming swampy areas was still used by the Aztecs of Mexico City at the time of the Spanish conquest. The surplus of food provided by this highly productive agriculture permitted others to build the great city complexes.

Atlantean society was ruled as always from the top down. Rulership was allied with the priesthood and both maintained their rule by a system of institutionalized fear; fear of the gods, fear of the rulers and fear of the priesthood. Ceremonies to appease the gods included as a necessary element the blood of human sacrifices. The sacrificial victims were either prisoners captured in war or citizens selected for the purpose. After each bloody execution, the bloodless body was then thrown down the temple steps, dismembered, and sold in the market for food. This form of religious cannibalism was still being practiced when Hernando de Cortez and his men discovered Mexico. Indeed, the Spaniards' desire to abolish this cannibalistic religion was their strongest justification for the eradication of Aztec Culture. Modern Mexicans deplore, as crimes, this cultural genocide but they fail the mention "the rest of the story."

The Aztecs however were not the immediate successors of the Atlanteans but a people who migrated into Mexico from the North at a far later time. Between these two periods lay an interval of perhaps thousands of years. The Aztecs were well aware that there were ancient cities buried in the jungles in the South, but they strictly avoided them. They feared to enter because they believed them haunted by the spirits of the "old ones." The Atlantean ruins were as foreign to them as they are to us today.

The art depicted on the Atlantean buildings both amaze and appall the visitor. Even as ruins, these buildings have a power of architectural order that remarkably satisfies our senses of proportion. Their architecture exhibits no rote repetition of a single style, copied from city to city, or even from adjacent buildings. Each building is unique. Each ruined city has its own set of motifs distinctive to that city, yet, having an underlying relationship to the styles of its neighbors.

In added upon the building's surfaces are decorations that completely cover their exteriors. These depict images and symbols utterly foreign to our minds. With the examples of ancient Egyptian, Babylonian or Scythian art we can empathize with these to some degree. These artistic works represent to us themes and ideas with which we feel an instinctive bond.

An ancient Eurasian artist's work may differ from our own but still be accessible to our instinctive understanding. Those artists are from *our* past. They had not our exact vision of life but, nevertheless, we "know" and "feel" what the Scythian artist meant to impart. They were *our* forbearers and *we are* their descendants. Our hearts and minds carry the felt echoes of their traditions and motifs.

But, concerning the art of the Atlanteans, there is nothing, no psychological link whatsoever to *our* culture. The images depicted in their art appear to us to be the work of interplanetary aliens. They exhibit stylized designs and forms hitherto unimagined. Their half-human half-animal figures indicate a creative sense gone wild. The ghouls found carved even in the living rocks are the faces of tormented demons. Strange figures peer out at us from the walls of every building. One such horrid face may be seen at the top of the staircase on the western face of the temple perched upon the pyramid at Uxmal; its gaping mouth forms the temple's door through which must have passed countless numbers of sacrificial victims. On The walls of other temples are images of their sacrificial victims at the instant of their decapitation—with their necks spraying out the last drops of their arterial blood. On other walls, are carved piled up rows of human skulls collected from the remains of countless victims.

The Atlantaens were obsessed with this rite of ritual murder. It was demanded by their cult of institutionalized Satanism. The people lived in a state of constant fear; fear of being chosen as their god's next victim. With Satan as their god, the world for them was a living Hell. Included in their panoply of sub-gods was the cult of the sacred snake whose image forms the frieze about many buildings; a tribute to the Devil's earthly form.

Their pottery and sculpture reinforce the conclusion that they were obsessed with demonism. Many human images are squat, ugly and fearsome. In asking oneself what these figures might symbolize, one senses they must depict the demons of Hell. Indeed, while we cannot read the hieroglyphics inscribed upon their buildings, the story of their religion and governmental system is writ plain in their arts and architecture. These show a society held in thrall by a royal priesthood that was dedicated to Satan and to his rites; by a religion celebrating evil for its own sake. Atlantean Culture merited no mercy and it received none; for it was destroyed on the first day of the Flood by the justified wrath of the living God.

When did the Flood occur? What was the date of that great Flood that destroyed Atlantis and its people? One piece of significant information is that the inhabitants of Central America were then quite familiar with the American Mammoth or Mastodon whom they seemed to have worshiped as divine because they carved their images upon their temples. (i.e. on the earliest Puuc temple buildings at Chichen-Itza and at Uxmal). Currently it is believed that the Mammoths became extinct at the end of the Pleistocene Era, approximately 10,000 years ago. (Encyclopedia Britannica 1970 Ed. Vol. 14. p. 1049) This means the ruins of the Atlantean cities are 10,000 years old—or older.

Supporting evidence as to the date of the Flood comes from a report by the Greek historian, Herodotus, who visited Egypt in the year 450 B.C. Being, himself, an initiate in the Greek Mysteries he was accepted by the Priests of the Egyptian Temple of Vulcan. When he asked them what was the age of Egyptian civilization they replied that they had kept a faithful record of their own priestly line of succession and that it extended back for 11,344 years! (History of Herodotus Tudor 1928 p. 131) Adding 450 years to this figure would mean that the founding of this Egyptian priesthood occurred in the year 11,794 B.C!

Might it not be that, following the apparent disappearance of Atlantis, the earliest Egyptians, actually, were an isolated Atlantean outpost, forced by what seemed to them to be the destruction of their homeland, left to carry on Atlantean civilization alone. Might this not explain the undoubted similarities between the earliest examples of the Egyptian's arts and architecture with those forms we find in Central America? Might it not ultimately found that the cultures of Egypt, Europe and Asia are not the "Old World," at all, but are rather the "New World"—the "New World" after the Flood?

CHAPTER FOUR

THE CHRISTIAN REVOLUTION

Repeatedly one hears these questions: "What is the cause of the general decline in our civil morality?" "Why are the world's political conditions becoming worse, rather than better? "Why does each social remedy fail to have a positive effect? Why is it that the more we pay for education the less the students learn? Why is it the more we pay for crime prevention the higher the crime rate? Is there a conspiracy at work to create these problems? As these social evils intensify, everyone is beginning to ask; "what really is going on?" If there are answers to these questions, what are they?

For those who ask such questions the reply is; yes there *are* answers to these questions but questioners must be warned that, upon receiving those answers, many will blanch in horror at the reality they expose. To those who argue the question as to whether our past and current history illustrates the workings a systematic conspiracy, with them no further discussion is possible. If dreamers must dream; let them dream on. If the majority wish to believe in the validity of the superficial political and social events that we see upon the surface of the waters, described as "the news," we can provide no answers. We speak only to those capable and serious minds who know, without a doubt, that, just below the surface, there *is* a secret force at work

What is this force that creates the wars nobody wants; that determines the outcome of elections and political events; that in the case of the obviously political assassinations of three American presidents, who or what organization prescribed the outcome of the subsequent investigations? How does it impose its intellectual prostitution upon the media, Congressmen, Supreme Court Justices and Presidents? And why *is* this force so concerned with prohibiting "Christian" prayers in public schools and in public places while it preaches toleration? Why is it so adamantly concerned with the rights of women while, in the same breath, it advocates the extermination of unborn women still in the womb? Why does it advance, as social liberation, every act of depravity while it tirelessly protects the rights of the perpetrators? If this is a conscious conspiracy, it must be motivated by the very Spirit of Evil and be the greatest possible enemy of mankind.

If there is such a force, who dares to speak its name? Few do. But a remnant must always speak out. Among these is the author who will speak its name; trusting in the power of Christ and in the living God for protection who states that this force is, at its core, the Synagogue of Satan; a diabolical army of slaves who wage war eternal upon the Kingdom of God and His people. This evil thing appears before us in a thousand guises. It poses as being Progressive political movements Orthodox religions, the "New Age" Cults and as the Fraternal Lodges of "brotherly love." Each of these pretends to be only a charitable organization, intent upon doing good works, but each has a common bond, the enshrinement of its "reason" over the Word of God.

They all agree upon one point. Universally, they deny Christ. These organizations ensnare the unwary into their nets by convincing them that submitting to their beliefs will insure personal wealth in this life and, yes, even godhood after death. They feed upon the basest instincts of human nature. Superficially, they deliver upon their promises. Their members sometimes do become the rich and famous—they may even become the politicians, priests and popular leaders of the day but at a terrible price; that same price that was exacted by Mephistopheles from Faust—the price of their immortal souls.

Yet, in this merciless war, the armies of Satan encounter a mightier enemy. It is, and always will be, the Lord Jesus Christ and His Gospel. Christ's Gospel convicts them. Daily, it burns their souls like fire.

Figuratively, Christ drew a line in the sand when He said; "you are either with me or against me." Either one submits to Christ and follows His commandments—or they do not. There is no middle ground. Christ declared eternal war upon Satan and his followers. Satan made the same offer to Christ in the wilderness as he later did to Faust; the offer of earthly rule and power. Yet unlike Faust, Christ refused. This Culture War still rages today. It is the war between the followers of Satan and the followers of Jesus Christ.

Yes this present writer is an "educated" man who once believed in the rational objections: How can anyone *prove* that Jesus of Nazareth, a man who lived in the year 33 AD and was crucified could arise from the dead on the third day? Scientifically, of course, it cannot be proven. For believers, faith is sufficient. While to skeptic minds there will never be proof enough. But the question of scientific proof is not the issue. The central issue is the ideational divinity of Christ. His divinity can be and *is proven* not through physical demonstration but by internal evidence contained within His teachings. These are the proof. Make this test. Obtain a red-letter edition of the King James Bible (if you can find a copy) with the words of Christ printed in red for easier reading. Read and seriously understand what Christ said. In that time or in ours the plain meaning of His

Gospel is the most revolutionary doctrine ever preached! The Gospel, itself, contains the proof you seek.

In that time and forever before, the world was ruled by fear and violence. People knew no other system. Attempts at representative government in Greece and Rome were short-lived and soon reverted back to the rule of force. Customs and laws were tribal or national. Moral actions were the observance of rules of conduct between members of one's own tribe or city-state. Toward outsiders the rule was; "do unto others before they do unto you!" Outsiders were "barbarians," not to be trusted—undeserving of mercy.

At this unlikely point in history Christ spoke and said: "Do unto others as you would have them do unto you." Today, most civilized people take this statement to be self-evident. Then, it was revolutionary. No one had ever viewed human relationships from this perspective. This single commandment is the very foundation of Western Civilization. It is the basis of what we call "Fair Play." "Fair Play" is fundamental to the rule of law. Without Fair Play, self-government fails and the tyranny of force and superstition replaces it. The naive assume that everyone agrees with this ideal but this is not the case. Under the rule of "rationalism" "humanism" "communism," or any other non-Christian culture, there is no observance, whatever, of this ideal. In those cultures the rule: is every person and class takes all they can get without regard to the rights of others. These acts they justify as self-interest; yes, but it is self-interest without quarter. In countries without Christ we see only unremitting confrontations: Jews against Arabs Muslims against Hindus, tribe against tribe, no toleration, no mercy. Those non-Christian philosophies explain why the world still writhes in unremitting agony.

"Fair Play" in practice means there must be one set of laws that must be applied in every case; without respect to persons.

Fair Play is the basis of games. To have a game, there must be rules; and the rules must be the same for all. Games are but rehearsals for life. Without rules, the game of football would become a gladiatorial contest—waged, not for points, but for the extermination of one side or the other.

Fair play has long been associated with the Christianized English character from earliest times. Once, during a Viking raid upon England in the 10th Century, a small army of English confronted a band of Vikings who, unfortunately, had been trapped upon a small peninsula surrounded on three sides by the sea. In the interests of "Fair Play" the English decided that the Vikings had no decent chance to fight from that narrow position, and so, they called out to them; inviting them onto the open shore so that both sides could fight fair as equals. This was a "beau geste" of great significance. Unfortunately, according to the accounts of the subsequent battle, the English were defeated and forced to pay tribute. But yet, this was of no great concern to the sporting English, because their *"rules of the game"* had been observed. (The Battle of Maldon, 991, AD)

Fair Play is the foundation of both English and American Common Law; an artifact which many nations pretend to imitate. But, when, as is usual, these fail to include within their laws the spirit of Fair Play, all else is meaningless. When their laws are twisted by lawyers or by their courts to give unfair advantage to one side or the other;—when winning, not justice, becomes the goal, their legal systems fail and become the weapons of tyranny.

In a world ruled only by force, Christ provided the certain path toward peace. He spoke and said: "Do unto others as you would have them do unto you." This commandment remains so philosophically elegant it provides the certain proof of Christ's divinity.

In another revolutionary teaching, Christ founded the science of psychology when He said: "In the eyes of God, justification is based, not upon the acts we commit outwardly, but upon those we commit inwardly." It is a sin to think of a sinful act even though we may not perform it. To imagine harm to an enemy is a sin. This commandment reveals the nature of sin. Sin is of the mind and sins of the mind are the cause of most mental problems. A sense of deep internal sin is often the cause of serious mental illness. Christ's teachings recognized the primary importance of our mental lives. Christ provides the only effective method of mental healing.

To heal the mind it is necessary to purify the mind—to convert the mind from the contemplation of evil to thoughts of righteousness. This is proven by the successes of Christian psychology. Alcoholics Anonymous and similar organizations find that only a religious conversion can provide a lasting cure. The success of Alcoholics Anonymous is founded not upon rational arguments over the evils of alcohol but upon the commitment of the victim to submit his heart and mind into the care of God. Until the alcoholic or the drug addict makes this submission, alcohol or his drug remain his god and neither Alcoholics Anonymous nor anyone else can cure him.

Christ's teachings that declared the primacy of the mental life over the physical, was an explosive concept in a world that, until then, believed only in the reality of the physical. His teachings revolutionized religious thought which up to that time was preoccupied with the practice of symbolic sacrifices and rituals to placate the wrath of their idols. Because Christ preached that each man and woman had their own conscience and was an individual before God, it followed that each man and woman had value—within themselves. This created the idea of "Individualism." If God recognizes each one of us as an individual then each of us must also be a separate entity from the state.

This was an undreamed of concept until proclaimed by Christ. It was then, and remains today, political dynamite. Continually, it resists the efforts of the social planners to treat mankind as a herd of animals. Individualism means that men have God-given and inalienable rights including the right to their private thoughts and persons and by derivation to have a voice in their own government. These ideas led to a belief in the rights of the citizen—later memorialized in the Magna Carta and, yet later, in the United States Declaration of Independence. Christ's commandments form the only viable basis for civilized government.

Christ made a further astounding pronouncement when He said: "All men are equal in the sight of God!" In an era when one man, Caesar, was counted as worth ten thousand ordinary mortals, this was heresy; but the idea spread like wildfire throughout the Roman world. Within three hundred years, this doctrine brought the Roman administrative system to a halt. As people, at large, became more convinced of the idea of social equality, the Empire weakened. Caesar and his officials were seen to be men just like themselves. Christians observed the public laws but ceased to believe in the tenets of the hierarchical Imperial system. Finally, in self-defense, the Empire of Rome was forced into a corner and into an accommodation with this new Christian philosophy.

Religiously, the idea that all men are equal before God created another revolution. It declared an end to the need for a priesthood upon which all religious practice had heretofore depended. If all were equal before God, there was no need for intercession by any human intermediary whether he be a Shaman, a Rabbi or a Priest. Christ became the High Priest whom all could approach in prayer. The priesthoods of the world's Old Religions suffered a mighty blow at the Christ's crucifixion; when, as it is written, "the veil of the temple was rent." The old order, based upon the mystery religions ended and the Kingdom of God came into being. Upon entering into God's Kingdom, the new believers found, for the first time in their miserable lives, the joy of spiritual peace and liberty.

Christ's divinely revealed teachings created the modern world. As a body of ideas they represented a new philosophical system that came into being without warning. No one at that time could have imagined such a set of ideas. They were revealed to us only by the wisdom and mercy of God. Christ's commandments created understanding, not hate, equality, not slavery, justice, not tyranny; and gave access for all to the throne of the Almighty God.

These revelations by Christ must, literally, have come from God because none of *us* on this planet, at any time, would have invented them. God sent his Son in the form of a man to announce His Kingdom to the world—to a world that otherwise would have remained locked within a static series of tyrannies and religious superstitions. Christ showed the Way to Salvation. That Way remains open today. These doctrines *prove* the divinity of Christ.

But the enemies of Christ did not rest. Since the Crucifixion they have conspired to reestablish the Old Order and the Old Religion which is Paganism allied with tyrannical government. That spiritual and political warfare against Christ is the root cause of all the wars and revolutions mankind has suffered during the last two thousand years. Since the crucifixion, and up to the present day, there has been a continuous war being waged against Christianity by the Anti-Christian powers. Their goal is to subvert Christ's gospel.

Those groups are legion. They are familiar organizations that many of us have joined as unwitting dupes. Who is there who has not been invited to be a member of an organization and been deceived by the argument that one should "join in" because it is all for a "good cause?" Our purpose is not to condemn theose individuals but to bring to their attention the false-doctrines and organized deceptions that enslave the victims of the Anti-Christ.

THE EMPIRE OF ROME

Knowing the generally good intentions of the Catholic laity, the present writer recognizes that most, in their hearts, desire to be moral and to do the will of God. Most believe that the embodiment of virtue is to be found within the Church and through the Mass. While they may be mistaken, the honesty of their faith will be rewarded. Perhaps, it is only the simple faith of Catholic believers that restrains the central organization to within any bounds, whatsoever. The believers do not see or have ever been told what transpires behind the scenes in the Vatican; as it directs the workings of the Roman Imperial System. The Roman Catholic Church can only be understood by looking back to the time of its inception when its character was originally formed.

After the Crucifixion and the Resurrection of Christ, and for the next 300 years, Primitive Christianity was an underground movement. Its practice, when discovered, could be a death warrant for the believer. The Roman's intolerance of Christianity was based upon their certain belief that it was politically subversive. Other sects could be tolerated because they acknowledged the divinity of the Emperor. Christians, however, would not worship the Roman gods or the divinity of Caesar. They obeyed the civil laws but otherwise looked to a higher authority.

Under the worst forms of persecution, this Church was proven in the fire. Necessarily, they were a band of true believers who followed the gospels as first delivered to them by the Apostles. They became a silent opposition and a growing embarrassment to the Roman State whose official religion was the worship of Jupiter (Zeus) who was served by a collegium of priests and Vestal Virgins who guarded the sacred fires among their other duties. So interlinked was the Roman government with this Pagan religion that it was thought necessary that every high official must have also served as a priest in the temples.

In addition to the Christian underground, the Empire had other problems. As the power of the state became centered in the person of one man, each succeeding Emperor faced the eminent possibility of being replaced by a coup d'etat. To rule was perilous. And, externally, Rome was in a continuous state of defensive war against the swarms of barbarians who were pressing in upon the frontiers from all sides. Luxury and sloth had weakened the Romans' will. The citizens of Rome, once, had been the only persons permitted to serve in the legions Now, they were too rich and too comfortable to fight. As a result, the Empire came to be defended by mercenary troops; recruited from the conquered provinces. These were unreliable at best—and, at worst, might overthrow the state. Ultimately, the rulers came to realize that, strategically, their situation had become untenable— that a general disaster loomed; that their moral and political center was giving way.

To solve this dilemma the Emperor Constantine and his administrators made one of the most fateful "command decisions" in world history. They determined to merge the Roman Empire into the strongest of the subversive religions— Christianity. Rome decided, in effect, that; "*If you cannot lick them, join them.*" In the year 312 AD, by the Edict of Milan, Constantine wedded the Roman state onto the Christian underground movement and, thereby, added to Rome's power the fervor of these new allies. By this single act, the Christian Church in Italy was taken over entire. In essence, it became the Roman Imperial Government concealed within the robes of religion. Thus began the Second Roman Empire; a.k.a The Roman Catholic Church

With the government's takeover of the Christian religion, what followed was remarkable to the Romans of that day. Within a few years their temples, dedicated to Jupiter, became empty of followers or were converted into Catholic Churches. Pagans such as Augustine became, first converts, then pillars of the new faith. It seemed to contemporary observers that the cult of Jupiter had just faded away. This was, of course, an illusion. What actually had taken place was the incorporation of the Cult of Jupiter within an outward shell of Christianity; it was Christian without but Pagan within. As proof of its Imperial status the new Church was given the signal mark of a "state religion"—the tax exemption.

Constantine, who had held the title of "Pontifex Maximus," under the old religion, now arrogated to himself this same title in the new Church. He became its first Pope. This new State Church was established upon the Pagan model so that once again it interposed itself between the laity and God. A new priesthood was formed who declared that only they could deal directly with God, take confessions and forgive sins—doctrines; lifted from the Pagan Roman system. The reestablishment of priestly rule, separating the people from their God, was a direct violation of the teachings of Christ; but was essential as a system of command and control upon the Imperial Roman model.

Outwardly, the political apparatus of Roman government, the Senate and its administration, began to wither away; but, effectively, they were replaced in every city and province by the newly minted priests of the Church. Overrun as it was by myriads of barbarians; and with raiding bands often camped at the city's very gates, Roman administrators found shelter within the Church. As their method of operation they invented a new plan for the conquest of the world. The world would be conquered and controlled, not by the power of Roman armies, but by psychological means; by using the power of religious superstition. The believing laity became an army far more powerful than the dispirited legions they replaced. From this time forward and for the next twelve hundred years the Imperial Catholic Church controlled both the religious and political elements in every part of Europe. A King might assert he ruled by divine right but the Pope, God's Vicar on Earth, dispensed that right at his pleasure. The Pope was superior to all. Upon the threat of an interdict, any King who failed to obey, was forced to come on his knees before the Pope in Rome to beg for mercy.

Although the Roman Imperial Church extended its rule throughout the old Empire, hidden in remote areas, true Christianity carried on. These communities still believed upon the Bible and followed the original Gospels of Christ and the teachings of the Apostles; especially those of Paul. They were sometimes called the Paulicans or Baptists. But, because they refused to foreswear the Gospel of Christ and submit to the Popes, they were accused of heresy. Persecutions repeatedly followed. Again and again these independent church communities were driven by military force from one place to another throughout Europe. Their sufferings were unimaginable yet they preserved the true faith while continuing to withdraw from the grasp of the Catholic authorities.

As their colonies would again begin to flourish, new Crusades were called down upon them. The Crusade against the Albegensian "Cathars" (the word means the Purified or Puritan) of Southern France, is well documented. (1209 AD) The Cathars' crime was that they believed they could not be saved by the "humanistic" Catholic religion and, as a result, set up their own churches; wanting to purify themselves before meeting judgment in the next world. To the Popes, this was a threat to their very being. Therefore, they ordered their Cardinals, Bishops, Priests and, finally, the French King, to wage a series of deadly Crusades against the Cathars; offering these Crusaders, as bribes, title to all the property of the Cathars whom they could murder or drive away. As an additional inducement to those who "took the Cross," the Pope gave them each a lifetime "*Indulgence*" absolving them of all their past or future sins. With that, the innocent Cathars were hunted down like dogs and when caught, burned alive. The Cathars' "heresy" was finally exterminated; but only after fifty years of repeated campaigns.

As late as the year 1572 the Protestant Huguenots of France were butchered in the merciless Saint Bartholomew's Day Massacre; carried out at the urgings of the Jesuits, the Pope, the Queen Mother, Catherine de Medicis, and by her son, the young King Charles IX. Once begun, this series of massacres continued throughout France for the next 30 days. Catholics in the provinces, following the example of their King, his mother, and of their local priests, ultimately killed upwards of 50,000 Huguenots. The succeeding king, Henry IV, a Huguenot sympathizer, as an apology, granted the remaining Huguenots equal religious and civil liberties by the terms of a proclamation he called his "Edict of Nantes." (1588). Nevertheless, one hundred years later, in 1685, at the urging of his Jesuit confessor, King Louis XIV revoked the Edict of Nantes—and, once again, deprived his Protestant subjects of all their property and of their civil and religious liberties. In this final "Holocaust," any Huguenot who would not abjure and return to Catholicism was forced either to emigrate or die. By this final act, France stripped itself of some 400,000 of its most diligent and loyal citizens. Many fled to America where some were the founders of New Rochelle, New York.

From the time of Constantine until the present day, little has changed. The Second Roman Empire (as the Roman Catholic Church) remains a continuation of the First Roman Empire; masquerading as a Christian religion. It confirms its pagan origin by its position that whenever a doctrine of scripture conflicts with the dogmas of the Church, the "Dogmas" prevail. The canon law of this Second Roman Empire remains, in the eyes of the Papacy, supreme. This explains its fundamental lack of principle in its doctrinal positions. It is not without reason that the Roman Church was deemed in the Bible as "The whore that sits upon the Seven Hills" for its guiding rule has been and is to accrete into itself any other doctrine or religious symbol that will protect it and foster the advancement of its central Imperial Political Organization.

Thus we see incorporated within it the Egyptian cult of Isis and Osirus. The Goddess is displayed as the Madonna figure with Osirus, (Christ), as the murdered son. The Madonna represents not only the physical mother of Jesus but she is ranked now as the "Mother of God!" If this outrageous title were not enough, the Pope has recently announced plans to issue a further decree—granting the Madonna another title; that of "Co-Redemptrix"—raising her, literally, to be a Goddess as the divine partner with her Son. Thus are Isis and Osirus to be fully incorporate into the Catholic panoply without shame.

Christ is denied by the Imperial Church; He being depicted, always, as crucified, defeated—dead upon the cross. In this way the power of Christ is shown to be ineffectual. He could not even save himself. Note that the prayers of the Rosary are addressed, not to Christ, but to His mother, the Madonna. Do these prayers not begin with the words ("Holy Mary Mother of God ..."). Yes, well we understand the poetry and the hypnotic effect of the Rosary—and how much and how sweetly Catholics love their Madonna—but they worship a vain thing.

To them I say there *is* good news. Christ was not defeated. Christ is not dead. Christ triumphed. He rose upon the third day and reigns now at the right hand of God. The Kingdom of God *is*. It was established at the Resurrection and will continue on forever. It exists *now*. All who hear the call may enter into it today. Those who live in this Kingdom are the body of true Christians who follow His Gospel as revealed in the Scriptures. They live in a parallel world—the Kingdom of God.

That the Imperial Church is an accretion of Pagan religions, posing as Christian, with the Roman Cult of Jupiter as its central element may be further illustrated by examining the signs and symbols it employs. The College of Cardinals were originally 70 in number—a continuation of the Roman Collegium who were also the 70. Even the color of their robes remains identical—cardinal red the color of Satan? Such congruencies cannot be accidental.

There appears also the use of idols (graven images) in every phase of the religious worship—this practice is expressly forbidden by the Second Commandment of God. The saying of prayers to the Madonna and to the bones of dead persons (alleged to be "saints"), rather than to the living God, is a denial of the First and Second Commandments. Outrageously, the Jesuits gloss over this violation by saying that since Christ appeared on Earth, in the flesh, the commandment against the making of graven images was revoked and is no longer in effect—a classic example of the Jesuits' lying casuistry!

To ensnare the unwary, the Jesuits have created a new list of Commandments that eliminates God's Second Commandment, prohibiting the making or the worshiping of graven images. In their new version, God's Second Commandment is deleted and replaced by a new Second Commandment. This new Commandment is only a rephrasing of the First Commandment. This is a most clever trick on their part. They still have Ten Commandments but the prohibition against the making and worshiping of statues and dried bodies of embalmed "saints" has now, most conveniently, been made to disappear.

Totally unrecognized, is the use in the regalia of the Catholic bishops and the pope of the symbols of Dagon the god of the Philistines. Dagon was held by them to be the greatest of all the gods because he existed in the form of a great fish and was therefore the only god to survive the Flood.

To honor him, his priests dressed themselves in costumes in the form of a fish with the open-mouthed head of the great fish upright; forming the cap or miter upon the head of the priest with the rest of his garment draping down to form the body of the fish and its tail. (Reverend Alexander Hislop "The Two Babylons" Loiseaux Brothers Press 1916 p. 215)

That the cult of Dagon lives on within the Roman Church is revealed by the shape of the miters worn by the Pope and his Bishops upon formal occasions. Their caps are a replica of the head of the fish god, Dagon, and although not one Catholic in ten million suspects the connection, the traditional instruction to Catholics to eat "fish on Friday" is to do honor to Dagon not to show contrition by foreswearing meat. Note: that the symbol of this fish god is now used by modern-day Catholics to identify one to another. Catholics assume this symbol represents the work of the apostle Peter "as the fisher of men" but in this they are mistaken.

These humanistic and Pagan elements undermine the moral thinking of Catholics everywhere. Taught that the priests can forgive sins, Catholics are subtly encouraged to satisfy their lusts because their sins can be washed away by confessing them to a priest. Each year at the festival of Fasnacht or Mardi Gras, that begins the forty days of Lent, the Church authorizes the celebration of a public orgy. During these revels, fornication, drunkenness, and gluttony are encouraged so as to satisfy the people's carnal lusts in preparation for the rigors of the coming period of denial.

In nations still ruled by constitutional laws as in the Unites States and Europe, Catholics are encouraged to look upon the local laws and customs as mere barriers to their personal freedom. The practical moral code of Catholicism may be summed up in the following statement: "Catholics are permitted by their Church to take and do whatever they can get away with!" In truth, "good Catholics" are, first and foremost, subjects of a foreign nation-state, the Vatican State in Rome; where the dictatorial Pope's authority is superior to all other national governments; including that of the United States! When the Vatican Nation State sends ambassadors to another country, these they call "Nuncios." This signifies that the ambassador is not a diplomatic negotiator but rather that they are "Announcers;" announcing the "infallible" positions of the Pope who, in fact, is Caesar; Emperor of the Second Roman Empire.

In its drive toward its One World Dis-Order the Church preaches the overall "inclusiveness" of all people races and nations without distinction. This explains their neutrality regarding interracial marriage and their favoring of the free emigration of peoples across international borders. It explains their toleration of homosexuality, lesbianism, and other forms of moral deviation. In the United States where it closely directs our government, our politicians are unable or unwilling to stem the invasion of our country by Mexicans, Arabs, Orientals and every other race and religion that legally or illegally enter into the U. S., daily, in their thousands. Those policies are designed to achieve the final goal of the Church (i.e. the Second Roman Empire), which is to create a polyglot single one-world human population. With this accomplished, they plan, by the use of eugenics, to breed themselves up as elites and to breed down a vast peasant population to serve them.

One of the most wicked of its acts is the demand that the priesthood submit to an oath of celibacy. Christ said. "Do not foreswear yourselves." "Swear not." Let your yea be yea and your nay be nay." What could be plainer?

The Church's requirement that their priests take an oath of celibacy is contrary to nature and to the Word of God. Taking this oath places the priest in an impossible emotional and physical situation. Once committed to this course, is it any wonder that most, under the demands of the flesh, lapse into secret fornications, adulteries, homosexuality, pederasty, or alcoholism. In practice, what the oath really means is that priests are only forbidden to legally marry. Otherwise, whatever they choose to do will be excused. The result of this oath is that celibacy drives the individual priest toward either sin or psychological breakdown. Celibacy was characterized by one priest as being "a factory for madness." It undermines the moral self-confidence of priests who now, from a sense of personal guilt, become unable to honestly believe in what they are supposed to teach. Most preach no more than the platitudes; limiting themselves to performing the deadening rituals of the office, but little more. Celibacy turns the priesthood into, guilt ridden robots who must follow, without question, the orders from above. The more honest and moral renounce their oath and leave the Church while the weak and cynical remain. The purpose of the Church is not to purify the priesthood but to pollute it; to enforce its discipline upon these now warped and isolated individuals, up and down the chain of command, who find they have no other course but to serve out their sentences, literally, as psychological prisoners.

CHURCH AND STATE

It is a given that politics and religion go hand in hand. The notion that there exists within any political system a complete separation between government and religion is utterly naive. Governments maintain their rule with the tacit assent of the population simply because the governors are less numerous than the governed. The method of gaining this assent has always been to attach to government an element of moral and religious legitimacy to ensure the people's loyalty. This was true in ancient Egypt it was so in Rome and obviously it is the situation in the Muslim and Jewish states today. Worldwide Jewry while it everywhere supports the cause of religious toleration for itself, practices almost none in its own state Israel. In Israel there is no toleration—for it is their law that only certified Jews may be allowed to immigrate into Israel and become citizens.

The reason modern Americans have so much difficulty with the question of "the separation of church and state" is because, in practice, it is impossible. All governments have a religious position even if it is only atheism. As a practical matter, every politician and official in government has a religious position. Catholics, Protestants, Jews, Moslems or Atheists do not shed their religious beliefs at the door of the Congress or at the door of any other office.

This unavoidable fact—that religion(s) and government are inseparable, was wisely addressed by the American Constitution that, in order to create a multi-religious government, resolved the problem by saying that the government would not take sides by financially sponsoring any religion or sect.

Psycho-political rule is the ultimate technique of governments—to rule not by visible force, excepting when necessary, but, rather, by psychological deception and persuasion. This being the case, the ruling party finds it expedient to work as a nameless secret organization to advance its policies through front organizations whose members appear to be the religious leaders, educators, politicians, journalists and civil servants. It is the Secret Party that determines, at any given time, what is politically acceptable and correct. It defines the "Party Line" on any issue. In fact, and in practice, it also creates opposition parties with slightly opposing "party lines" in order to maintain the appearance of there being political diversity.

From time to time this force issues its religio-political manifestos that advance its goals. Some describe these goals as the "New Age." Others speak of the existence of a "Grand Design" although not revealing the names of the "Grand Designers."

Because the organization seems so faceless, it is difficult to discuss and to quantify. Therefore we will assign it a name: "PAGAN." PAGAN is the combination of Anti-Christian forces that are attempting to rule the world. It must rule all. After establishing its power in any area; thereafter, it governs by a system of social and political thought-control; tolerating no opposition. The Presidents, administrative office holders, governors, city mayors, councilmen, newspaper and magazine editors, movie producers and television and radio station broadcasters serve only at its pleasure. Since PAGAN became established, what we think of as the "news" is usually only a series of synthetic events. The protest marches, the abortion controversies, the ethnic conflicts and the terrorist acts—almost all are fostered and funded by PAGAN.

The real news is—there is no real "news!" There is only the replaying of previously effective scenarios designed to advance their purposes. PAGAN writes history—in advance! Their scenarios are then performed in the *"theatre of the real"* with each massacre, agitation, political movement, and bloody revolution further advancing its plans.

During the sixty-year conflict between the Arabs and the Jews in Palestine, whenever a settlement appears to be possible, a new bombing or atrocity occurs; perpetuating their mutual hatreds. The "troubles" between the Catholics and the Protestants in Northern Ireland are no nearer to being resolved after eighty years of massacres. Any attempt at a settlement between the reasonable elements on both sides is blocked by new acts of terrorism. Peace? "'tis the last thing wanted."

In all regimes and, yes, under our American system as well, there is an extreme emphasis placed upon state control of the means of communication. The postal system, telephones, radio broadcasting, and publishing invariably are either state controlled or closely licensed so that PAGAN may monitor all interactions between its citizens. The current installation of a SIM chip in every new cell phone to facilitate official wiretapping lets the Government know the physical location of every user. The request for a "key" to permit the interception and decoding of privately encrypted E-mail is another recent example. PAGAN must know all—PAGAN must control all.

The most visible of the secret societies set up by PAGAN to carry out its plans are the Lodges of Freemasonry along with their dependant organizations such as The Lions, the Rotarians etc. Masonry is a worldwide organization with millions of members who, for personal advantage, place loyalty to the ideals of Masonry before any other moral or religious commitment.

WHERE AND WHEN DID FREEMASONRY ORIGINATE?

Masonic chroniclers assert that, following the suppression of the Knights Templar by King Phillip IV of France in 1312, some fugitive Templars found refuge in Scotland. Later they say, the organization spread southward into England and from thence to around the world.

Whatever truth there may be in this story, the first public Masonic Lodge was not constituted until 1717 but there had been underground precursors. One member was the famous antiquarian and prominent government official Elias Ashmole, founder of the Ashmolean Museum at the University of Oxford, in England. It is said he was initiated into Freemasonry in 1646 indicating that Masonry was operative then. There is little doubt that from the time of the founding of the Templars at Jerusalem in 1150, secret societies of Templars and Rosicrucians existed among the upper classes of Europe and Britain. Literary works of the time such as "The Roman de la Rose," "The Morte d'Arthur," and the "Parzival Legend," covertly taught their secret doctrines. Those groups privately incubated until political conditions required their expansion. That is the official story.

Masons boast that Masonry is but "the pagan religions revived" and this certainly describes its spirit. But what solid evidence is there for the theory of its ancient origin or are these tales a fable designed to "*conceal rather than to reveal?*"

"Freethinkers" in the Eighteenth Century who organized themselves as Masons to advance the ideals of logic and reason probably believed that by denying Christ they were, thereby, opposing the superstitions of the Roman Catholic Church. (This was the theme of Mozart's opera, "The Magic Flute.") But, in this belief, they may have been greatly mistaken—taken in by an even greater deception.

Understanding that the Imperial Roman Church was built upon the foundations of the Old Roman Empire and that it opposes, always, any revival of true biblically based Christianity; was this new secret society the enemy of Rome or was it a wholly owned subsidiary? Since both denied the doctrines of Christ, how could they have been enemies? Those who believe themselves capable of deceiving the Vatican will find they play against the very masters of deceit. The Papacy invented the principles of the dialectic long before Hegel, Marx or Engels. The first rule of Realpolitik is that since there must always be two sides to every question the best method for maintaining autocratic rule is to create and to control both.

The medieval Order of Knights Templar was an independent and financially successful international trading and banking organization; at first tolerated by the Church. But then, the Templars had to be suppressed because they became so very wealthy they threatened to overshadow both the Papacy and the civil governments as well. Still considered a potential threat, the Order of Knights Templar has never been permitted to revive. Templarism was not the precursor of Freemasonry.

From internal evidence it is almost certain that what we now know as modern "Accepted Masonry" (meaning its members are "accepted" while not actually involved in the craft of stone-masonry) is, in point of fact, the Jesuitical System of the Catholic Imperial Church; covered over with a facade of Paganism, Rosicrucianism and Templarism; formed into a militant organization aimed at the subversion of the Biblical Christian faith. Freemasonry is a sub-branch of Jesuitism designed to entice the common run of ambitious but unprincipled non-Catholic men and women who, for the offer of social companionship, financial advantage, and/or political preferment, swear terrible oaths of loyalty to an organization whose origins and ultimate purposes they will never comprehend.

Consider the workings of the Society of Jesus; the name itself being a blasphemous lie, for there is nothing in Jesuitical doctrine or practice that supports the Gospel of Christ. Rather the Society of Jesus that was formed in 1539 by the soldier of fortune, Ignatius Loyola, was, from its beginning, the militant secret service of the Papacy. Its avowed aim was and still is to defend and to advance the power of the Second Roman Empire and its Popes—by whatever means are necessary. It is the Vatican's Gestapo! (Encyclopedia Britannica. 11th Edition. 1911 Vol. XV. P 327)

The Order was given the blessing of Pope Paul III, the then current pope. He approved special rules that released the Jesuits from the religious practices governing other orders. Then, the Pope absolved them, in advance, for any crimes they might commit in pursuit of their common goals.

At that time, the Church was faced with its greatest challenge since its founding; the Protestant Reformation then spreading throughout Europe caused by the printing of the Bible. Having the Bible permitted people for the first time to read the Gospels of Christ. It revealed to them the true nature of the Papacy and many Catholics began to follow the teachings of scripture rather than the made up traditions of the Church. Millions left the Church and became Protestants. This, the Church deemed heresy. To oppose this movement, the Jesuits took up the role of intellectual leadership; becoming the Inquisitors the educators and the political advisors to the Catholic kings and aristocrats. In carrying out these roles they grew into a powerful *political* force; both respected and feared throughout the world.

During the Sixteenth Century, in the minds of the Popes, their greatest loss had been the conversion of England to Protestantism under King Henry the Eighth, in 1545. During the remainder of that century it was the fixed goal of Jesuit intrigue to bring Great Britain back under Catholic rule. Its greatest attempt to achieve that goal was their assault upon England using the Spanish Armada, in 1588. The Armada's purpose was not to engage the British Navy but to avoid it and to rather land a Spanish army in England; to capture Queen Elizabeth and to cut off her head. Afterwards, they planned to reestablish the Catholic monarchy. Miraculously, they failed in this attempt. The defeat of the Armada, by Sir Francis Drake, was the great turning point in modern history; for it meant that the Reformation throughout Europe was spared.

After the natural death of Elizabeth in 1603, the Jesuits formed yet another plot to set off a political revolution in England whereby they might reestablish the Roman Church as the official Church and to restrict or prohibit Protestantism. Under the direction of underground Jesuit priests, a group of lay Catholics; (Guy Fawkes and 20 others), planted a mine of gunpowder under the House of Parliament; intending to assassinate King James I along with most of its members and to, thereby, destroy the Protestant Government of England at a single stroke. Providentially, the plot was discovered in time to prevent the explosion. The conspiracy was entirely the work of the Jesuits' and never was supported to any degree by the lay Catholics of England. But thereafter, fear of the Jesuits and of Popish plotting, caused deep concern about Catholic loyalty and for the next two hundred years, under the Test Acts, Catholics were prohibited from holding any public office in Great Britain—a proscription that lasted until 1829.

The autocratic and Catholic-leanings of the next king, Charles I, formed the root cause of the English Civil War. (1641-45). At issue was the question of the "right" of the king, under his divine appointment, to rule his country by decree without consulting the desires of Parliament. But the Civil War and the triumph of Cromwell's Puritan Army of "Ironsides" again blocked the efforts of the Jesuits to reincorporate Great Britain into the Church of Rome. Afterwards, when Charles' son Charles II, was restored to the throne in 1660—while he remained a lukewarm Catholic, he agreed to maintain England, officially, as a Protestant country. However, when upon his death, his younger brother, James II, a more militant Catholic, became king he violated this condition and tried to reestablish the Catholic Church as the official Church of England. This violation of his coronation oath so exasperated the Protestant English that, by common consent, he was deposed into exile in 1688. Next, his daughter Mary, along with her Protestant husband William III, Prince of Orange, were invited over from Holland to become the joint King and Queen of England. Two years later James II, returning from exile, raised up a new Catholic army in Ireland for yet another attempt to regain England for the Papacy by force of arms. The effort failed when he was defeated at the Battle of the Boyne in 1690 by King William III's army. That ended the Jesuits' efforts to conquer Great Britain by military means.

Confounded but not deterred, the Jesuits returned once more. This time disguised as "wise men" or "wizards" they slipped back into England and brought with them a new message. It was that there was a new, enlightened, organization that had discovered secret knowledge of the ancient mysteries of the East—knowledge that would confer great benefits upon its members. To those who joined, that knowledge would reveal, even to the least of them, the way to power and riches. Within the organization everyone would be equal. Peasants and tradesmen could hope to associate with their lords upon equal terms—a great attraction to the lower classes.

While mysticism, like magic, in its various forms had existed in Britain as a scholar's private hobby for hundreds of years, its amalgamation into its present system did not occur until 1717 when the founders of Accepted Masonry met in a London tavern to organize themselves into the "Grand Mother Lodge;" the Lodge from which all other Masonic Lodges have since derived their authority.

Proof of the Jesuitical foundation of Masonry is demonstrated by the fact that this first Lodge immediately adopted the Jesuit's system of organization. Membership was to be held by men only and in a series of three degrees in the same manner as in the Society of Jesus. The three grades in Masonry are: Entered

Apprentice, Fellow Craft, and Master Mason, identical to the Society of Jesus. In Jesuitism the brothers are ranked as: Novices, Scholastics, and Temporal Co-adjudacators. This confirms that what we call modern Freemasonry is, in reality, an amalgam of Jesuitism and Paganism organized for the subversion of Biblical Christianity. Note, also, that since founding Freemasonry, the Jesuits have withdrawn from direct political activity and have made no further attempts to interfere in British political life. There is no need. Their Masonic dupes now perform those tasks to perfection. It is by this means, by using Freemasonry as an underground army, that PAGAN has gained mastery over all Western governments. As an arm of the Jesuits (PAGAN), Freemasonry is the largest underground political force in the world today.

This secret network, along with its subsidiaries, comprises most of the true governments of Great Britain the United States and of every major nation in the world. The organization works toward regional and finally a one-world government based upon its religion of atheism. The evidence confirms this. In the United States Masonic emblems are arrogantly displayed on many public buildings and even upon our currency. The Masonic symbol of the 13 stepped pyramid surmounted by the all-seeing eye of Osirius may be seen upon the reverse of every United States one-dollar Federal Reserve Note. This symbol was placed there by the decree of President Franklin D. Roosevelt in 1933.

Clear minds must face hard facts. A full understanding of history confirms that the American Revolution was the product of a Masonic conspiracy. Consider those who were its founders. Ask who made up the Committees of Correspondence? In point of fact, they were the Masonic Lodges, North and South, who operated under this pseudonym. The revered founders of the American Republic were nearly all Masons; with Benjamin Franklin at their head. He was Grand Master of the Colonies.

It must be undoubted that the United States has been ruled by Masonic politicians from that day to this. You may ask what about the two-party system? What about elections and what about "democracy?" Are these not our protections against tyranny? The answer is no! It is an act of folly to believe in such chimeras. What triumphant political power would ever permit a real opposition to exist? The answer is never! Were there actually to be two opposing parties soon there would remain only one; with the other physically eliminated from the scene. During and after the American Revolution the true opposition party were the Tories; those who remained loyal to the King. It is estimated these constituted about one third of the population. After the Revolution, all the prominent Tories were arrested and their property seized. Then they were deported as penniless refugees to Canada. So much for toleration and for the two party system!

How does this master-party carry on its rule without appearing to exist? It is a simple plan, repeated over and over in the world of the "democracies." The single, Masonic Party, divides itself into two sub-parties. One posing as Conservative, allied to tradition, and the other as Progressive, open to change. They are called Federalists and Republicans, Conservatives and Labour, Republicans and Democrats, Right vs. Left etc. etc.... ad infinitum. Choose your country, choose your party labels. It is the same everywhere.

During the recent charade carried out in the Soviet Union, the single Communist (i.e. Masonic) Party, upon orders from PAGAN, split itself into two so that Russia could present itself as being modern and democratic. That this far-rago was swallowed whole by the majority of educated Americans and Europeans is amazing. The citizens of Russia know better. They know that little has changed. In truth, the people of the Soviet Union took no part in any revolution. The so-called "Glasnost revolution" was simply an arranged political entertainment because PAGAN had decided to change the rules of the game. As the hapless prisoners of the regime, the Soviet people found that one day their jailers had unlocked their cell doors. This was and remains to them, bewildering. They find it difficult to believe that, somehow, they have been given any degree of personal freedom. Many still miss their cells. This same performance had previously been rehearsed. and acted out. in the freedom movements of Czechoslovakia, Poland, Hungary, Bulgaria, and Rumania; in none of which did the people set *themselves* free. This process has been called the end of the "Cold War"; but it was a "Cold War" that had never existed except as a propaganda device employed by the world-controllers to hold their subject peoples, both East and West, in a state of perpetual mobilization against fictional enemies.

For a full understanding see Machiavelli's "The Prince," or Arthur Koestler's "Darkness at Noon," or George Orwell's "1984," They describe the process completely.

CHAPTER FIVE

THE GENEALOGY OF THE ANTI-CHRIST

Christians must relearn the lessons of their past and recognize the enemy they face. They must study the motivating ideas that lie behind the record of events. While all cultures pass through organic stages of development from birth to fulfillment and to eventual decline; what differentiates them is that each is the carrier of a unique set of intellectual and religious potentials. Parallels may be drawn showing that Greek music developed by logical stages; steps that were repeated sequentially during the Faustian Culture in Western Europe. Spengler describes these parallels between cultural developments but stops short of the final goal which would have been an exploration of the question; why do paralleled cultures produce such differing outcomes? Spengler failed to see or feared to say that the reason for these differences is religious—or that religion determines destiny.

The culture of Ancient Greece cannot be understood without recognizing the effects of its mystery religions that were centered upon the temple at Delphi and its cult of the Python. This temple of the Python was built over a volcanic vent from which noxious gases were continually fuming. According to myth, these vapors were the breath of The God Apollo, appearing as a gigantic snake which lived below in the cavern. Peasants and Kings alike made pilgrimages to the temple to bring with them rich offerings in hopes of having their futures foretold and to beg for guidance that would insure their success in politics or war. During her séances, Pythia, the Pythoness, sat upon a metal tripod-seat built over the smoking vent. As she became intoxicated by the fumes, she would fall into a trance-like state and was able, supposedly, to interpret the will of Apollo. In practical terms her advice came to control the leaders of the city-states; from behind this veil of superstition. Orders given by the Oracle became the controlling factor in the politics of ancient Greece. By her advice given, now to one city-state and then to another, the Oracle determined, in advance the direction of political events. One honored member of this order was Pythagoras i.e. "He who has the wisdom of the Python."

Is the Oracle of Delphi dead and forgotten? No; not at all. Every second year a torch is lighted from its eternal flame by the current Pythoness and it is then relayed across the world by long distance runners to light the flame that burns throughout each meeting of the Winter or Summer Olympic Games. Indeed Apollo and his Oracle are alive and well today.

The motifs of our present culture are still religious but with one fundamental difference. Ours is a culture with two religions locked in a struggle for mastery. One is the Kingdom of God and the Gospel of His Son Jesus Christ and the other is the Religion of Paganism; collectively called the Anti-Christ. Forever opposed, these define our culture today.

Christians need not despair. Being saved, they live in the Kingdom of God and have nothing to fear. Paul said "If God be with us who can be against us?" But to oppose this enemy we must first outline his order of battle:

First there was Satan.

Satan founded the Pagan religions to separate the people from their God. He provided them with a myriad of mistaken beliefs. He created, as their gods, a belief in the divinity of animals, or of idols or the worship of their own physical members; symbols of the generative principle. These false gods demanded human sacrifice even unto the sacrifice of the people's "first fruits;" their own firstborn children. These inventions of the Devil were and still are the "Old Religions" the fane of Paganism.

Defeated by the Resurrection, the followers of the Anti-Christ renewed their struggle by the only means left to them; secrecy and subversion. The Roman cult of Jupiter (Zeus-Apollo) merged itself into the Primitive Christian Church and begat the Second Roman Empire.

The Second Roman Empire begat the Roman Catholic Church and its Priesthood.

The Catholic Priesthood begat the Jesuits.

The Jesuits begat Freemasonry.

The Jesuits begat the Illuminati

Freemasonry begat the American Revolution

Freemasonry and the Jesuits begat the French Revolution.

Freemasonry begat the United States "two party system."

Freemasonry begat Mormonism

Freemasonry and the Jesuits incited the American Civil War.

Freemasonry and Zionism financed Karl Marx and his Communist Manifesto.

Freemasonry begat B'nai B'rith.

B'nai B'rith begat the Anti-Defamation League (ADL)

The Anti-Defamation League begat the American Civil Liberties Union (ACLU)

Freemasonry begat the theory of Evolution.

The Papacy incited World War I.

The Papacy incited World War II.

Freemasonry and Zionism created the Russian Revolution.

Freemasonry and Zionism directed the "Soviet Experiment" for 70 years.

Freemasonry created the United Nations.

Freemasonry fabricated the fake "Cold War."

Freemasonry begat the Liberal Protestant Denominations.

Freemasons resurrected the Arian Heresy.

Freemasonry resurrected the doctrine of Premillenialism.

Freemasonry "revises" the King James Bible to destroy the Gospel of Christ.

On and on ad infinitum.....

A MESSAGE TO THE JEWS

PAGAN, a worldwide cabal, entices into itself any and all religious or political groups willing to ally themselves against Christ and his Gospel. Jewry which thinks it cannot be deceived in truth *is* being deceived. Within Judaism PAGAN has created an atheistic front organization called Zionism. Zionism's founding principle is that of Machiavelli who taught that in order to form and maintain a state, it is necessary to provide the people with an enemy; real or imaginary.

To provide that enemy, the rabbis and Talmudists have for hundreds of years perpetuated the myth of outside oppression. To preserve the identity of the Jewish tribe, it maintains, as fac,t that the people and government of every host country, whether Christian or Moslem, are their natural enemies who without warning may commence another pogrom against them. Therefore they reason, to preserve their tribal identity, it is necessary, as a matter of course, for Jews to subvert the existing religious and governmental system in each host country. Zionism itself does not observe the morals and customs of these countries but opposes them with the plotting of a subversive fifth-column. Zionism employs the notions of "us" and "them;" ideas that explain the phenomenon that in whatever culture they find themselves, Jews do not assimilate into the local society but remain by, choice, a "separate people."

These are characteristic of the Jewish psyche. They produce an attitude of political and social divisiveness; which in most cultures finally results in a violent reaction. Jewish alienation is the essence of the "Jewish Problem." A "Jewish Problem" does exist but it is a problem created by the Zionists and not by their neighbors. If and when an irruption does occur, wherein the feared pogrom becomes a reality, and whereby thousands or millions of Jewish innocents are massacred; secretly the leaders of Zionism rejoice. They celebrate because their predictions of doom have once again come true and in their view the suffering of the "little Jews" only purifies the Nation and makes it stronger. As an example while untold numbers of ordinary religious Jews were left to suffer and die in the Nazi pogroms, the Zionists had long-since escaped to shelter in the West. Zionism now uses its endless retelling of these martyrdoms to justify and strengthen Zionism's ongoing political agenda.

A recent account of the effects of Zionist alienation upon one Jewish family and upon three generations of Jewish immigrants to the United States is provided in a recent autobiography written by David Horowitz, the one-time editor of "Ramparts Magazine," the voice of the "Progressive New Left," at the University of California at Berkeley in the 1960s. His book is a "mea culpa" a confession of guilt, an attempt to expatiate himself from the wickedness he fostered as a leader in the New Left Progressive movement. (Horowitz, David "Radical Son." The Free Press New York 1997.)

A portion of his story describes the lives of his father and mother who were both second generation Jewish-Americans; born of immigrant parents who had come from the Ukraine in 1905. While both his father and mother had had excellent opportunities provided by nearly free American college educations, they both nevertheless joined the New York Communist Party underground in the late 1920s when that Party's avowed aim was to overthrow America by a violent revolution.

Throughout their lives these "true believers" never wavered in their radical beliefs. Even in their declining years they still asked their son, David, "when do you think the Revolution will begin?" So committed were they to the Communist Ideal that although both were well paid as teachers in the New York Public School System they adamantly refused to buy the rented house they lived in; even when it was offered to them for only $8,000.00. To them, the ownership of property was a sin. They preferred to rent!

David, their son, absorbed these beliefs and after his graduation from Columbia University in 1959, he took on the role of a free-lance journalist and spokesperson of the Progressive New Left (i.e. revolutionary communism) in Europe and in Great Britain. Returning to the U. S. with a wife and young family he next settled in Oakland California to be near the University of California's Berkeley campus because he had been hired as the editor of Ramparts Magazine. For the next few years his articles in Ramparts fueled the fires of the student demonstrations at UC/Berkley.

As the student rebellion peaked, Horowitz began also to assist the Black Panthers organization in Oakland, California. The Panthers were led by the murderous criminals; Huey Newton and Bobby Seal. As their foolish mentor, Horowitz raised money that was supposed to be used, he thought, to advance the African-American side of the revolution. But, in so doing, he failed to recognize what and with whom he was dealing until, finally, the Panthers murdered his female associate who had failed to follow their orders. This revealed that the Panthers were not an idealistic left wing black political movement but rather that they were a pack of drug-crazed gangsters. Horowitz found himself face to face with primal fear and with the smell of his own approaching assassination; a thing he had not bargained for.

A more revealing fact he discovered was that one of his friends, a staff writer on the magazine, had been and probably still was an undercover agent for the U. S. National Security Agency (NSA); an espionage agency so secret, at the time, that few politicians in the Federal Government even knew of its existence. With this, Horowitz began to suspect the real truth; that he and Ramparts Magazine were not in the forefront in any real protest movement against the government but that they were, in actuality, dupes of that same government; set up as a controlled opposition.

Thus Horowitz discovered PAGAN and found that he was not a fighter for freedom on the leading edge of a "revolution" but that he was just one more cog on the wheel. He saw that his whole life up to that time, and the lives of his father and mother before him, and the lives of the whole Radical Left for the prior seventy years had all been wasted fighting in a fake revolutionary war from the wrong side.

The image of the "Wandering Jew" forever seeking his homeland, may otherwise be viewed as an intellectual "wandering" as the Jew searches for his Messiah even while that Messiah might already exist before his unseeing eyes. A prominent Rabbi was once asked "Suppose it turns out that Jesus of Nazareth actually *was* the Messiah?" And this rabbi replied "Well I voulden't be a bit surprised!"

PAGAN

PAGAN that cabal of anti-Christian forces made up of Jesuits, Freemasonry, Zionism and every other non-Christian organization has been the determining force behind political events worldwide for the past two thousand years. As such it must be held responsible for the monstrous tragedies its deliberate tactics have brought about.

But the line in the sand remains. Christ's followers stand on one side; His enemies on the other.

A "Letter to the Editor" recently printed in a small town newspaper expresses it thus:

"There are two world views—the Christian view and the non-Christian view. These two views are in total antithesis to each other. They are totally different concepts and conclusions about morality, society, government and law."

"The polarization of our communities our nation and the world," the author continues, "is between the Christians and the non-Christians. There is no way to mix these world views. They are opposed entities and cannot be synthesized or compromised." (Author's name withheld)

Lionel Trilling, the late English literary critic who, at one time taught at Columbia University, defined the program of PAGAN thus; "The entire purpose of the *modernist project* (i.e. atheistic revolutionary socialism) is the complete overturning of two thousand years of Judeo-Christian tradition which aims to turn every virtue into a sin and every sin into a virtue." What could be plainer?

A PERSONAL NOTE

By exposing the nature of these religious and secular institutions, the present writer hopes to assist the reader in passing from darkness into the light. Yet none may claim superiority over another. Have we not all at one time been snared into a religious or secular organization whose purpose was to deceive? Prior to receiving the Gospel of Christ the writer labored for years as a "true believer" within an organization whose stated purpose was to promote improvements by education and political action; but whose real purpose was to divert and to neutralize our efforts. If I offend please forgive. If you cannot forgive I must accept the consequences and take the position of Martin Luther who when he stood on trial for his beliefs before the Holy Roman Emperor, Charles V. and the assembled Catholic Inquisition at Worms, said; "Ich kann nicht anders" ("I cannot do otherwise.")

Seek the truth
Listen to the truth
Teach the truth
Love the truth
Abide by the truth
Defend the truth
Unto death.

Prayer by Jan Huss

THE LEGACY OF ROME

By its first act the New Imperial Church reestablished the pagan priesthood that once again separated the people from their God. When they were told they were now Christians the ex-Pagan believers found themselves in very familiar surroundings as they entered the newly minted Catholic Churches.

Its second act was to withhold the scriptures from the laity who were thereafter provided only with priestly interpretations.

Its third act was to encourage immorality among the laity. If the priesthood could forgive sins, sin was thereby encouraged. One's sins could now be erased by confessing them to a parish priest who claimed he was a direct agent of God and whose favor might be purchased for a price. Until the Reformation, Papal Indulgences were being regularly bought and sold that could pardon sins for periods of months or years. It was claimed that the Pope had a "treasury of indulgences" from which he could sell to forgive the people's sins at his pleasure.

Its fourth act was to reintroduce such Pagan elements as the elevation of the Madonna and the Christ Child as veiled representations of Isis and Osirius which is idol worship and by making the statues and relics of dead people objects of veneration. These are not inconsequential matters. They deny Christ and the Sovereignty of God. A Catholic when beginning a journey says a prayer not to God but to St. Christopher. If he becomes ill he is directed to consult with St. Jude. And finally when he dies, Christ is still denied. Christ's power to save the soul of a believer is counted as insufficient. No it is the final rite of "extreme unction" that must be administered only by a priest that supposedly will absolve the dying Catholic of most of their sins. But even then, the souls of the dead must still suffer in Purgatory until released by further prayers said by the priests and nuns who extract a fee.

Its fifth act was the policy of establishing the Imperial Catholic Church everywhere by force. Under strict Catholic rule, and in those countries where the Church had political control, dissent was illegal and the liberty of conscience was heresy. As late as 1550, in England under Queen Mary, Protestant "heretics" were still being burned alive at the stake. Has this system of punishment been changed or is it merely being held in abeyance until it can be reapplied?

A sixth act was the failure to maintain the purity of the faith. Rather in every land where the Catholic religion has been introduced they have not so much converted the subject populations; as they have amalgamated Catholicism with the endemic Pagan religions; thus creating a panoply of "other gospels." This is now officially advanced as the policy of "inculturalization."

A seventh and most monstrous act is the institution of the ceremony of the Mass. Having been present through a high mass held in the English language and while attempting to understand the symbolism of what was being performed I became horrified at the import of the ceremony. Under the Doctrine of Transubstantiation the Church insists that the wine and the bread are converted into the literal body and blood of Christ. If that be so then the Mass is a re-crucifixion of Christ performed over-and-over each day.

After preparing the elements for Communion, the priest turns from the congregation toward the back alter and holds up the urn of "blood" and says "let this sacrifice be acceptable unto you." To whom is he speaking? To God who could not wish Christ's sacrifice to be repeated? Or is it to some other god who would welcome Christ's blood as his gift?

After the "body and blood" are consecrated they are offered to the celebrants to *eat and drink!* This beyond any doubt is ritualistic cannibalism. The participants in this ceremony are ritually murdering Jesus Christ and then literally eating his body and drinking his blood in company with the god of the Catholic Church, Satan.

The Church has the effrontery to call this ceremony "The Sacrifice of the Mass." It is contrived by the twisted logic of this pagan Church that Christ as the "Host" willingly offers himself, His very Body and His Blood, up to be re-crucified daily for the benefit of feeding the faithful. It must be seen then that the meaning of the Mass is a continuation of the Pagan belief that God requires the sacrifice of a human victim; even though God in the example of Abraham and his son Isaac specifically taught that human sacrifice was henceforth not required and was forbidden.

Contemplation of these acts is appalling. For those who might wonder if there is such a thing such as a "Black Mass" they need seek no further. The "Black Mass" is the one performed each day in every Catholic Church.

Please Catholic friends come out from such a thing and out from these Pagan rites. When you enter a Christian church you will find the cross empty—signifying that Christ is risen, not defeated, and that He now sits at the right hand of God—presiding eternally over His Kingdom and His people.

THE KINGDOM OF THE CULTS

Other so-called religions masquerade as Christian but are in actuality cults. Eager to employ the name of Christ; they deny His divinity. A cult is any group or teaching that adds to or modifies or twists the Gospel of Christ by claiming to have received a "new revelation."

This from the Apostle Paul: "As we said before *so* say I now again; If any *man* preach any other gospel unto you than that ye have received let him be accursed." (Galations I, 9)

"The Church of Jesus Christ of Latter Day Saints" is one such cult that bases itself upon a "new gospel" invented by the mountebank, Joseph Smith.

MORMONISM

Mormonism calls itself "The Church of Jesus Christ of Latter Day Saints" but it is in reality a, mock-Christian, Masonic Church. It was founded by one Joseph Smith who as a 14 year old farm boy in upper New York State, in the year 1820, claimed to have met Christ and Jehovah in the woods near his home. Some years later, he said he was visited by an angel named Moroni who told him where to dig up a set of golden plates upon which were hieroglyphic writings. Supposedly, these contained the story of an ancient North American race who were in reality one of the lost tribes of Israel. This lost tribe had been visited and taught by Christ in a second appearance on Earth. This is the origin of the name "Latter Day Saints."

Smith said he was visited again by Moroni who now lent him a pair of magic spectacles. These permitted him to read the plates and to translate them from the ancient Egyptian language into English. This "translation" is known today as "The Book of Mormon." Smith accomplished his "translation" during a period of months while working at his kitchen table. Conveniently, the angel told Smith to rebury the plates and to return the "magic spectacles" so that the "Book of Mormon," now, is the only primary source of Mormon doctrine. By concocting these scriptures, Smith so impressed the "rustics gathered round" that soon he became a local celebrity and as the group about him grew, it developed into the Mormon Church.

How had Smith been inspired to write the Book of Mormon? The theme may well have been plagiarized from a little known work by another author. Because, at this time, there was a great deal of interest about who were the builders of the Indian mounds then being excavated in many states in the Northeast, an author in Cincinnati, Ohio wrote a fictional work in the manner that we would call today "fantasy fiction." Its story was that the builders of the mysterious mounds were in reality one of the Lost Tribes of Israel.

Americans in the early Nineteenth Century did not believe that the American Indians they knew would have been capable of building such mounds. Smith, too, gave his story a biblical cast by making his work to be the history of a lost Jewish tribe. It was this work that Smith probably read and copied except in his version he employed the antique language and style of the King James Bible to give "The Book of Mormon" more authority.

Because, during this same period he had joined a Masonic Lodge, he grafted onto his new religion many of the symbols and practices of Masonry with which he was now familiar. These remain in Mormonism today. Most revealing is the fact that the Masonic symbol of the Compass and Square is still worn on the underclothing of all good Mormons as a protective talisman. Also, that in the Mormon Temples, during the most important ceremonies, the men must wear their sheepskin "aprons;" confirming that Mormonism is Mystic Masonry.

One central doctrine of Mormonism is the assertion that God was only a man who grew to become a God; so therefore any male Mormon may, by The Degrees of Perfection, also become a god.

Supposedly a "Mormon truism" is contained in the following couplet:

> "As man now is God once was;
> As God now is man may be."

("The Vision or The Degrees of Glory" N. B. Lundwall Publisher Salt Lake City no date p. 151)

This blasphemy is carried even further by their assertion that after death good male Mormons will be caught up into the Celestial Realm where they will be assigned to rule other Earths; empowered to make new Creations and to rule their individual planets as new sovereign gods-Jehovah!

To a Biblical Christian these ideas seem outrageous but they are *believed* by Mormons and represent a dangerous form of nonsense. It explains the powerful hold Mormonism has over its members. Not only are they to be saved but, after death, better yet, they are to be transformed into co-equals with God. This recalls the admonition in Psalms-2 that in their disobedience; "the people will imagine a vain thing."

As in Freemasonry, the Mormon hierarchy maintains firm control over the minds of its members. Discipline is enforced by inspectors who visit the homes of members each week. In a practical sense Mormonism is quite successful. It plays upon the fact that people without any other guide will welcome being included in an organization that gives them a program to follow. As a result of this attitude, Mormons are widely employed in the U. S. Government; many in high positions. Mormonism is growing because it preys upon the ignorance of simple minds but it remains what it always has been a Masonic cult.

B'NAI B'RITH
"The Sons of the Covenant"

Jews are nothing if not self-assured. Essentially they believe they are more hard working clever intelligent and farseeing than others; in point of fact the "Chosen People" or perhaps even the "Master Race." Yet they live as strangers in every land—feeling unease. Always the ear must be cocked listening for the approach of the next pogrom. Welcomed everywhere at first, they fear to trust. This fear continues in large part because never are they permitted to forget the past so as to make a new beginning. It is a fact of realpolitik that to create a nation it is necessary to create an enemy. An imaginary enemy cements the Nation by dividing it from its neighbors.

In Jewish teaching from the time of the Babylonian Talmud to the present day Jewish scholarship has emphasized the differences between Jews and Christians and has falsely taught that Christians have been the authors of all their sufferings when nothing could be further from the truth.

Pogroms there have been, but these were carried out by atheists; the leaders of the false religions. The Nazi persecutions were carried out not by German Bible believing Christians but by the Jesuitical National Socialists. The official religion of the NAZI government was "Deutsch Glaube" the belief in Germany; not Christianity. The ordinary Biblical Christians of Germany were as incapable of opposing that government as were the Jews.

How can any individual or group challenge a modern state by open rebellion? Yes the leaders of the main-line Protestant denominations or within the Catholic Church did little to oppose the persecutions but, as we now know, most of these had little to do with true Christianity. With common Christians it was another matter. When they could help they did; often at great personal risk.

As a companion book to the "Diary of Anne Frank" I would ask Jews to read another book: "The Hiding Place" by Corrie Ten Boom. (Ten Boom Corrie; "The Hiding Place" Bantam Books 1974.) Here was a Christian maiden lady who lived with her father and extended family in the City of Haarlem in the Netherlands. As Biblical Christians they hid young Jews fleeing from the German authorities at great peril to themselves. To do this they converted a room in their old house into a secret hiding place for young Jews on their underground road to freedom. This way they saved many.

Betrayed to the Nazis, she and her eighty-year old father were imprisoned and sent into the camps. Her father was never seen again. As those who lived in Europe in that time know, the camps were not for Jews alone. They held anyone opposed to the state; German, Dutch, and French, Russians, anti-Nazis, gypsies, homosexuals and whomever. Miraculously Miss Ten Boom was released alive at the end of the war and afterward spent the rest of her life telling her story; working for peace and reconciliation between all peoples.

In the same vein the people of Denmark followed the example of their King, King Christian, by wearing, en mass, the yellow Star of David in defiance of the Nazis. Whatever crimes governments may have been guilty of, the notion of "group guilt" should be foresworn by any intelligent person; whether such "group guilt" is applied to the postwar Germans or as it was applied to Jewry by the Nazis or as the Arabs and the Jews apply it to each other today.

It is a weakness of intelligent peoples to believe they cannot be deceived. The methods of PAGAN set every class and group against another; every religion against another so as to rule all sides by psychological techniques in the resulting confusion. To that end every type of deception has been practiced against the Jews.

In defiance to the Word of Jehovah, Jewish thinkers have been led into the mystic depths of Talmudism, Kaballah and other cultic beliefs all, all of them, teachings of the Devil. In modern times such a one is B'nai B'rith that many do not realize is Jewish Masonry: i.e. Jesuitism. It is Masonry for Jews only, yes, but otherwise it employs all the signs, grips and paraphernalia of Freemasonry. It was founded and first organized, in 1843, in a Masonic Lodge in New York City. Today, it has lodges in all the developed nations. B'nai B'rith conceals itself behind a facade of charitable and educational causes but it works, not for the advancement of legitimate Jewish interests, but for the purposes of alienation; causing it to oppose any and all public displays of Christian morality and practice. In all of these acts and movements Jews are fatally mistaken because, in all this world, bible-believing Christians are their only potential friends.

What does B'nai B'rith teach the Jewish people? It teaches that their enemies are the Christians, Christian moral codes and Christian influence in the state. In the United States, and around the world, Zionist intellectualism aligns itself with anti-Christian forces; from its love affair with Communism to its alliance with every other atheistic "liberal" or left-leaning movement from the days of Marx to the present time. Having been taught to oppose Christian influence in public education or in the state they accept the error that the U. S. Constitution decrees the separation of Church and State.

In the practice of law, Jewish attorneys rush pro-bono to the defense of notorious criminals so long as that particular defendant is a member of an "oppressed" class. The logic for this effort being the unstated belief that as long as these so-called "oppressed" can be protected, even when guilty, attention will be drawn away from the Jewish minority. In this they are again mistaken because by joining in these socially destructive activities Jews only aid and abet their future immolation. Should the Christian moral basis of our legal system be suborned, to the point it becomes ineffective or abandoned, this will usher in a totalitarian legal system based upon a state religion of atheism. When this occurs—no longer needed as PAGAN'S puppets—Jewry's turn will come again. Doubt it not for one moment; the true purpose of PAGAN is the elimination of *all* opposing religious groups; Christians, Jews and Moslems alike.

In a recent rehearsal, the massacre of the Seventh Day Adventist sect in Waco Texas that occurred on April 19th, 1993, PAGAN'S Forces invaded the State of Texas and carried out their murderous will without any shred of legality. This was a wake-up call for all religious communities. This assault inaugurated a new pogrom to be directed against any and all religious groups who do not politically submit; and the Adventists' leader, David Koresh, would not submit. On that day, PAGAN'S velvet glove of concealment was withdrawn to expose its fist of steel. Its next target could just as well be a Kibbutz or a Yeshiva.

What is to be done? First, Jews should withdraw from these agencies of the Devil. Then they should fall upon their faces before the living God and pray for His mercy and for a cleansing of their hearts and souls. With that accomplished, perhaps we all could all come together with respect and understanding, one for another.

A MESSAGE TO ISLAM

Moslems have long nurtured resentment against the Americans and Western Europeans whom they consider to be of the same breed as the "Franks" who invaded their lands during the Crusades. In their opinion, those invasions were without provocation. In the last half of the 20th Century after what they consider to be another invasion by the Zionist Jews, those resentments have been fanned anew into white heat. In this state of mind they equate Christians, Jews and Westerners together in one body as being their common enemy. But Islamic scholars should take a closer look at history wherein they would discover that it was not the simple Christians of Europe that made war upon them during the Crusades but, rather, it was the Second Roman Empire and the Popes who proclaimed Crusades against them. Little has changed. It is still the Second Roman Empire that directs its sub-governments and political front groups to war against them now. The Papacy and the Zionists do not know the meaning of the word "peace." It is only bible believing Protestants who, without the use of force, remain eager to bring Christ's Gospel to Islamic countries. It is the only hope for peace.

THE LIBERAL CHURCHES

After founding Freemasonry, the Jesuits directed it to assault Protestant Christianity by infiltrating their denominations and subverting their doctrines. Sowing the seeds of doubt in the schools of divinity, they revived the doctrine of Arianism a philosophy condemned by the Council Niciea in 327 AD. This humanistic view sets human reason above the spiritual mystery of Christ's atoning death and resurrection. Simply stated it argues that the biblical accounts of Christ's ministry should not be taken as literally true but merely as an allegorical story. While God may be omnipotent Christ, as a created being, must be less than God and in fact must have been no more than a man; a great teacher, perhaps, but not divine.

This doctrine was advanced again in England and in Europe in the mid Eighteenth Century during the brief period of civil and intellectual order that had been created by the Protestant reformation during the previous two centuries. In a relatively peaceful world this theory seemed logical to the idle intellectuals of the day. Arianism developed into an ordered system of biblical revisionism that produced the cult of Deism. The essence of Deism is that while there must have been a creator of the universe, once created, the Earth was left to run on by itself like a fully wound-up clock. God was finished, and all that has happened since has just been the result of cause and effect.

It is but one step more to the assumption that at this time on Earth there is no God watching and judging what we do. Therefore, all is allowed. Men are merely thinking animals, anyway, and whatever they imagine to do *is* natural. Nothing, per se, is either good or evil.

Eliminating God by logical analysis, these savants next found the need for a replacement belief and so they set up a new standard of moral order; Humanism, the self-worship of Mankind and of Mankind's Reason. They, thus, accomplished, in their fevered minds, the destruction of the Holy Trinity: Father, Son and Holy Ghost!

Deism lives on today and is the basis of Unitarianism and Congregationalism and is the core doctrinal belief of all the "liberal" denominations. Deism is Masonry masked with the cloak of religion. Its Supreme Being is identical with that "Great Architect of the Universe;" the god who is honored in the Lodges. As Deists deny every point in the Gospels, the Great Architect whom they honor must be the one who was thrown down from heaven; the Prince of Darkness.

A prime example of Deism in action was the French Revolution during which a prostitute was enthroned in the Church of the Madeline in Paris as the Goddess of Reason. The essence of that revolution was stated in its motto; "Liberte,' Egalitie,' Fraternite'." The words still echo in our ears. But what does the word "Liberte" mean; liberty from what? Fundamentally, what is meant is; freedom from the restraints of God's commandments; freedom to write and rewrite the moral codes to suit one's personal convenience.

"Liberty" as our courts now interpret it is the freedom to say or do *anything* without concern for the consequences. It is the freedom to commit pornography, blasphemy, libel, adultery, pederasty, murder or homosexuality—whatever.

"Liberty" allows all. This lusting after "Liberty" is not a new thing. John Milton judged this desire for absolute "liberty" four hundred years ago:

> "(they) still revolt when truth would set them free
> License they mean when they cry libertie."

(John Milton. Sonnet XII. "The Student's Milton." Appelton Century. 1933. p33.)

PREMILLENIALISM

Pre-millennialism is a perverse denial of Christ's divinity injected by Freemasonry into the body of the "fundamentalist" Christian churches who erroneously believe they stand four-square for Christ. The essence of the Pre-millennial doctrine is that Christ failed in His first coming to Earth. And, although He *was* the promised Messiah, He was rejected by the Jews and was crucified. He had promised to set up His Earthly Kingdom but this was prevented by the crucifixion. To remedy this defeat, Christ must return a second time to make a new war upon Satan and his legions and conquer them in a physical battle at Armageddon. *Then,* He will set up His Kingdom on Earth that will last for a final 1000 years. It follows that, at this time, we are living *before* the millennial reign and hence *this period is "pre-millennial."*

Pre-millennialists assert that after Christ completes His thousand-year reign, then, will come the "end of the Earth." The corollary to this story is the commonly held belief that because Christ failed at His first coming, Satan rules today; and the existence of present-day evil and wickedness is inevitable. Thus, the only course for Christians to follow is to endure for the present because, when Christ does return, they will then have their innings. This doctrine is propped up by excerpting ideas from the Book of Revelations the prophetic, last, and most questionable book in the Bible; a book that fails to agree in spirit with what Christ preached to His disciples.

The Book of Revelations is apocryphal and should be removed from the canon. Internal evidence clearly shows that it was written not by the apostle John but by some Byzantine author in the Fourth Century AD. It is filled with all manner of magical elements: "seals" "signs" "omens" "prophecies" and "mysteries" which are in fact and spirit wholly unrelated to the clear and easily understood teachings found in Christ's Gospels in the books of the four Apostles and in Paul's later Epistles. This "Book of Revelations" has been employed to found all manner of heresies that welcome the belief in a future physical battle must be fought at Armageddon to establish Christ's_*physical kingdom*. The Book of Revelation denies Christ's central message in which He specifically states that He came to establish, not a physical, but a spiritual Kingdom; a Kingdom, as He said, "not of this world."

These infusions of Arianism and Pre-millenialism into the Protestant denominations have led to a denial of the saving power of Christ. While Christ said He had saved all whom the Father had given Him, Pre-millennialism asserts He failed in that mission. It follows they say if the believer is to be saved at all he must save *himself* by taking some personal action. This becomes salvation by works. If he is Catholic he must attend Mass regularly, give enough money, and say his prayers to the dead saints and to their idols. If he is a Protestant, he must repeatedly come down the aisle to make his decision and re-decisions for Christ. These fundamental errors cause both Christians and Catholics alike to doubt the certainty of their salvation. In their hearts they doubt that what they have done will prove sufficient. Did they donate enough money? Were their prayers accepted? Can one moment of repentance cover over a lifetime of sinning? These doubts undermine the commitment of Christians. The Christian is told he must act to save himself but he knows in his heart that is impossible. These are the poisonous doctrines of Satan whose lies in the Garden of Eden sowed the first seeds of doubt in the minds of Adam and Eve—leading them to question a commandment of God.

Pre-millennialism explains the "touchy-feely" positions held by the liberal Christian churches wherein Christ and the principle doctrines of the New Testament have almost been forgotten. "Love" and submissiveness are the order of the day while biblical answers to society's problems are ignored. The laity come like sheep an-hungering for answers to their spiritual emptiness but these "hollow-men" these ministers of liberalism, having lost their own way, are powerless to feed them.

Their answer to the world's problems is always the merging of the denominations, with the schmoozing over of differences, until their final goal is reached— Ecumenism; which will be, in reality, the re-establishment of the Universal New Roman Imperium; a Religio-Political Empire.

That is the bad news. These are the "other doctrines" preached to us by the enemies of Christ. The good news is: because we now see these doctrines are the lies of Satan, the reverse must be true. Salvation of the believers is not just possible; no it is certain. It has already been accomplished; accomplished not by the miserable works of imperfect men but by the faith of Christ. As it is written because we all have sinned in our hearts, we are incapable of saving ourselves. But Christ, and the grace of God, have saved us. In advance, Christ has saved all whom God has "called."

Who are those who have been called? Are you among that number? The answer is yes you are. How so? Because, as you must understand by now it is not by chance that you have come this far and read these lines. Hear and Follow. Step over into His Kingdom today. This *is* the "Good News." This is the Gospel of Jesus Christ. May glory be unto God.

CHAPTER SIX

HISTORIC PARALLELS

Accepting Spengler's view that cultures are organic and pass through identical stages of development, where on the time scale is our culture today? Spengler's answer is in the title of his masterwork "Der Untergang des Abendlandes" ("The Decline of the West") wherein he argues not that Western Culture is heading for an immediate collapse but that it has fulfilled its genetic potential and is no longer "becoming" but is "become." Spengler's name for this final stage is "Civilization" marked by cosmopolitanism and skepticism; a period wherein any further advancement in the arts music or political themes can no longer be accomplished.

In the outlook of our time, artistic questions come to be seen as less and less relevant. To define greatness in music we are forced to look backward two hundred years to Bach Hayden Mozart and Beethoven. Our great period in painting and sculpture is now seen to have been the Renaissance and the Baroque. Distrusting new ideas in political thought we look back to ideas formulated in the Eighteenth Century. We look back to the principles of the Roman Republic as revived by the Eighteenth Century philosophers such as Locke, Rousseau, Hamilton Jefferson, and Washington. We honor the "founding fathers" as our surest guides in law and social organization. We find ourselves bereft of better ideas. Each new law in the United States must still be squared against the letter and intent of the Constitution of 1787. Tampering with this document is popularly viewed with great suspicion.

Spengler's conclusions continue to be confirmed. While beautiful and pleasing symphonic music was still produced during the 19th Century these works remain within the previously established melodic form they are variation and extension but not a further advancement. By the early 20th Century even imitation becomes difficult until by the end of the 20th Century no one is left who can even imitate the "masters." The audiences for "classical" music declines as succeeding generations lose their ability to empathize with its forms. The clangor of rock music now satisfies the needs of a new mass audience.

In the plastic and visual arts while imitation within the forms continues to be carried on the only "advancements" are the well-know prostitutions of art found necessary to serve the needs of the political revolution. Today's non-representational painters claim to express their inner selves and they do. Their blank canvases perfectly reveal the content of their minds. Expressing the skepsis of our age, they know nothing, they believe nothing and they paint nothing. Yet, this meaningless art is applauded to the skies by dealers and collectors. When artists do attempt to advance their work beyond the traditional forms—they cannot. For us, all possible forms have been fulfilled. Now is the death of art.

In politics a similar process has occurred. While lip service is piously given to the constitutional ideals set forth within the sacred documents there has been a "revolution within the form." The United States has progressed from a Republic into a state of bureaucratic socialism defined as "Democracy:" But as the belief in "Democracy" fades the next stage must be "Caesarism." Now comes the Empire; PAGAN'S rule from behind the screen of "Democracy."

The political history of the United States is best understood by examining the almost identical sequence of events in the story of Ancient Rome. Originally modeled after the ideals of the early Roman Republic, ours is a government of laws based upon principles that we assume to be unchanging. But like the Roman model, over time, it has grown from representative self-government into its present state; an administrative Empire. So alike, are the stages of their parallel developments, that even the basic political issues and the motivations of the participants are identical. Often, in making comparisons one need only substitute the names of the political actors.

Many methods employed by modern nation-states are derivations of the Roman System. Those principles live on today in the demand for legality and the rule according to written laws. In both regimes there is a civil service to create these laws and a standing army to enforce them. The system is familiar but it is adoptive. It is not of our own invention.

The Early Republic of Rome was governed by a Senate whose members originally were the humble farmers of Latium. These later came to be men appointed from the aristocratic and wealthier members of the community. Once admitted, membership was for life. Executive power was invested in two Consuls selected annually by the Senate from its most gifted members. These ruled with the full authority of the Senate but were appointed to serve for only one year. To be named a Consul was a great honor and a heavy responsibility.

Below the Senate was the Order of Tribunes whose role was advisory. They could propose laws but could not legislate themselves. They formed a body similar to our House of Representatives. Below these was the Equestrian Order made up of commoners, second sons of the aristocracy, and men of worth whose occupations could be commerce or administration but who were excluded from the

Senate or Tribunate. Military experience was considered vital and was a prerequisite to appointments to high office

Originally, this government functioned well, but the very success of Roman military expansionism strained it to the breaking point. Rome became a predator state living upon what could be extracted from the conquered provinces. The system was so lucrative that the citizens of Rome paid no taxes and were even supported by a new political invention—welfare. This came in the form of free food drink and entertainment. (Panem et Circensis)

The conquest of neighboring countries was made possible by Rome's organizational ability and by the natural free-enterprise economy that provided Rome with superior technology. As in our own time, communications controlled the ability to rule. Rome's network of paved roads throughout the Empire made good communications possible. Up and down these roads went the legions and their supply trains. Messages to and from the frontiers traveled quickly in both directions. After conquering a new province, the civil administration could rule from the center of power. The roads permitted the Legions to concentrate rapidly at danger points along the frontiers. Under the Republic private enterprise and inventions were welcomed not regulated and improvements in technology were rapid. These new technologies created a military force and a civil engineering system that for centuries was invincible. Rome was the world's first great Superpower.

Success however brought unforeseen problems. The incorporation of Greece and the Near Eastern provinces into the Empire fostered the importation of many foreign religious cults. Among these were the Phrygian Cybele, (Magna Deum Mater) (The Great Mother of God!), the cult of Dionysus (Bacchus), the Greek philosophies of Stoicism and Epicurianism, the Eleusinian Mysteries, the cults of Isis and Osirus, of Attis and of Astrology. The growth of these cults commenced the destruction of the Romans' belief in their ancestral family traditions and in the household gods which had heretofore unified the state. These cults undermined Roman standards and caused a great loss of respect for the values of family and common morality. They fostered an increase in divorce and a growth of luxurious excess. The "Decline of the Roman Empire" began at this time; proving once again that religion is the controlling factor in a culture.

Simultaneously, slaves were imported from the conquered provinces to be used as cheap farm labor. They provided cheaper labor costs to the industrial farms and displaced the yeoman tenant farmers who then appealed to the Senate for justice; but who went unanswered.

As the displaced workers sought employment, they drifted from the country into the City and swelled the numbers of the disenfranchised. Encouraged by ambitious Tribunes, these elements began asking for a voice in the government.

Thus began the rise of "Democracy" which inevitably proceeded on into "mobocracy;" to finally end in tyranny. Some Tribunes thought to lead and to appease this movement.

In response to the unrest there arose two Tribunes, the Gracchus brothers, who like the Kennedy brothers of our day, challenged the power of the establishment. In the year 133 BC, as Tribune, Tiberius Gracchus proposed a land reform bill to redistribute vacant lands to the displaced farmers. The Senate resisted and when Tiberius took his cause to the people and stood for election for an unprecedented second term, he and 300 of his followers were massacred on orders from the Senate.

A few years later his brother Gaius Gracchus, after also being elected a Tribune, revived his brother's program; proposing in addition the granting of Roman citizenship to the people of the allied states in Italy. These proposals had strong popular support but the Senate felt personally threatened. Once again it reacted. It issued an order to the Consul called "The Final Decree." This declared that the nation was threatened and it ordered the Consul L. Opitimus to defend the State by virtue of his *Empire*. This he did, by taking to arms and executing Gaius and hundreds of his followers without trial. From this point forward, it was obvious that changes in the rulership of the state would only be settled by force. The end of the Republic was in sight.

A parallel in our time are the Kennedy brothers John and Robert who may have believed themselves to be rich enough and popular enough to challenge the orders of PAGAN. As with the Gracchi brothers, they paid for this error with their lives.

While we are still close enough in time to these modern Gracchi to be aware of their personal weaknesses, it is likely that because they bought their way to power with their own money they were not directly the creatures of PAGAN. No doubt assurances were given by the Kennedy brothers that the advice of PAGAN would be heeded; but once in power they set out upon their own courses. They became "loose cannons"; that had to be put over the side to save the ship.

Recall those days in the fall of 1963. What were the major decisions facing President John F. Kennedy? First, there was the question of authorizing the grain shipments to a Soviet Union that was facing a winter of starvation without our help. The President had refused to authorize the release of this grain naively believing that the Soviet Russia was our enemy when of course it is a client state of PAGAN. Second, was the question of escalating our military intervention into Viet Nam. President Kennedy, who realized this policy would lead to an expanded war the nation did not need, refused the orders from PAGAN to do so. Like the Roman Senate when faced with a refusal to obey, PAGAN issued a "Final Decree;" an order that was quickly carried out by secret military forces.

President Kennedy's successor, Lyndon Johnson, was a man who, on the contrary, knew when to follow orders. It is rumored that on very the day of the Kennedy assassination, after being sworn in as president and returning to Washington, he made a long distance telephone call to the socialist-atheist guru and Harvard Professor, John Kenneth Galbraith. This figure was a leader in PAGAN. Was the reason for the new President's desperate call his need to determine the "party line?"

Whatever his instructions were, the new President called Congress immediately into emergency session before Christmas and forced it to pass the bill authorizing the required Russian grain shipments. As to the military escalation in Viet Nam he obeyed PAGAN to the letter. After this, the full force of the United States was unleashed upon that region. A ten year's long "police action" (officially not a war) was carried out entirely for the purposes of PAGAN without the informed consent of the American people nor of their Congress.

What *were* the purposes of PAGAN in Vietnam? The first was to pulverize the Buddhist religious culture of Southeast Asia because the Buddhists were stubbornly refusing to adopt Catholicism under pressures from their President Diem and his brother the Catholic Archbishop of Viet Nam. Second, the military operations were conducted as, real-time, maneuvers to experiment with newly developed tactical and logistical methods of making war (i.e. by vertical envelopment using helicopters) and to learn how to conduct counter-insurgency operations should there actually be a revolt anywhere on the planet. Third, It was to demonstrate to the world's leaders the overwhelming power at the disposal of PAGAN; military power that can and will be used against any state in the future that fails to cooperate. Examples are Panama, Granada, Iraq, Serbia, Afghanistan etc. From the standpoint of PAGAN the "war-games" in Viet Nam were a complete success and fully achieved their programmed goals.

In 1968 a second Gracchus brother, Robert Kennedy, took up the crusade by making his own run for the presidency. Like his brother, John F, he ignored the interests of PAGAN. It is necessary in our mythical two party system that the parties must hold the Presidency more or less by turns and this was Richard Nixon's and the Republican Party's appointed year. (President Johnson was ordered not to run and he withdrew in advance) But it seemed this might not be a Republican year. The populist appeal of Robert Kennedy was even more powerful than that of his brother and he became the front-runner by winning many of the Democratic primaries. On the day he was assassinated, he had just won the California Primary making it a near-certainty that had he lived he would have won both the Democratic Party's nomination and the Presidency as well.

Faced with another "loose cannon" PAGAN reacted even more swiftly; stopping Bobby Kennedy's rebellion dead in its tracks. The Democratic Convention of that year was then forced to nominate Senator Walter Mondale a non-viable candidate so that, as planned, Richard Nixon could be elected by a landslide.

In 1972 a third Gracchus brother, Ted Kennedy, was next offered the crown in the form of the Democratic Party's nomination; but he had seen enough. He replied to his supporters that not only would he not be a candidate but that he and his family would remain on vacation in Ireland over the rest of that summer. This Gracchus brother now understood the game and chose to follow orders. As a result he has enjoyed a long and happy career in the U.S. Senate.

CIVIL WARS

The attempted revolts of the Gracchi revealed a growing inability of the Roman Senate to control the cities mobs. It recognized that maintenance of civil order would depend more and more upon the use of force. This culminated in the rule of Sulla the most powerful of the generals whose army besieged and captured a rebellious Rome in 90 BC. To reestablish order, he assumed the title of temporary Dictator. He then resigned and retired; hoping he had safely restored the Republic.

Meanwhile the smaller surrounding states in Italy, allied to Rome, asked for the rights of Roman citizenship. When this was denied them, they seceded from Rome and attempted to form a new nation called "Italia" and they began a fight for independence; in a civil war known to history as the Social War.

Like the U. S. Southern Confederacy these rebels set up their own Senate minted their own coinage and soon had 100,000 soldiers in the field. Rome replied with its full might; waging a three years long war against this "rebellion." After four years, in 86 B.C., the secession was crushed; ending any further assertion of "states rights." However, to make peace the Senate was forced to grant the rebellious states the desired Roman citizenship with some restrictions.

In the United States a similar sequence of events occurred. The successes of free enterprise system led to a natural movement away from a Republic of totally independent states toward the establishment of a strong central government.

Because, from 1830 onward, the Southern States resented Northern intervention in what they considered local matters, the issues of "nullification" and of "states rights" arose. PAGAN decided that the ideas of "states rights" and of a voluntary Union of States must now be eradicated. They determined to settle the question once and for all by fomenting a war between the states.

Historians tell and retell in minute detail stories of this epic struggle but fail to provide us with the religious and cultural motivations that lay behind the American Civil War. They fail to explain *why* this war was waged. Certainly it was not waged over the slavery question; an issue that could have been resolved by negotiation to free the slaves and to compensate slave holders for their monetary losses. Such a program would have provided the freed slaves an orderly transition from slavery to freedom and it would have been far less expensive for the United States to have purchased and freed every slave in America than to have waged one year of the Civil War.

Historians tell of the two sides equally, matched in valor, that fought each other to exhaustion during those four long years from 1861 to 1865. The names of their battles still ring in our ears: Shiloh, Vicksburg, Chattanooga, Antietam Gettysburg, The Wilderness, Cold Harbor, Petersburg and, in finality, Appomattox. Historians glorify the great captains: Robert E. Lee, Longstreet Stonewall Jackson, Steven T. Early, Ulysses S. Grant, W. T. Sherman and Phil Sheridan. We are told all, except the final key, that would explain the reason *why?* The answer to this question is that, however they recount the deeds of these military figures, both past and modern historians fail to recognize, or hesitate to reveal, that, one and all, these leaders, and the two nations as well, were duped by a grand conspiracy carried out by secret forces they could not then comprehend.

Whether believing in their cause or merely seeking personal glory they all, like marionettes, danced upon the strings manipulated by PAGAN. In the North the forces of PAGAN: the Freemasons, the Congregationalists, the Unitarians and their forum the Union League Club of New York City, manufactured propaganda against the South; magnifying the slavery issue to enlist popular support. (But behind these were the Jesuits setting policy)

In the South, the Freemasons, acting through their special branches such as the "Knights of the Golden Circle," agitated for the right of the Southern States to secede. It was no accident that the war began in that hotbed of Southern Masonry; the City of Charleston, South Carolina; that was and still is the headquarters for the Supreme Council of the Thirty-third Degree of America and of the world! (But behind all these were the Jesuits setting policy)

Then, to make the waging of a full-scale war, even possible, one-half of the professional United States Army and Navy Officer Corps, were excused and directed to organize and to lead the Confederate Army and Navy; while the other one-half remained to organize and lead the Union Army and Navy. For PAGAN, it was like casting actors for a play; for an A-Team and for a B-Team; except that, this play was to be acted out in the "theatre of the real," with 400,000 Americans losing their lives.

It was Masonic hands that lit the first cannon's fuse when shots were fired upon Fort Sumter in April 1861. Thereafter, high Masons on both sides directed the progress of the war; repeatedly avoiding opportunities for a Union victory in order to continue the reduction of the Southern people.

During the Peninsula Campaign on June 1 1862, a Federal Army under General McClellan advanced to within four miles of Richmond, Virginia; the Confederate capitol. Its capture could have ended the rebellion, then. But for the conspirators this was too soon. If peace came now the higher goal—the total reduction of the South could not have been achieved. The South had not suffered enough.

What occurred next seemed inexplicable to contemporary observers. Under orders, McClellan dawdled before Richmond until General Lee took command of the Army of Northern Virginia. Then claiming he was outnumbered McClellan withdrew his army to the coast from whence it was evacuated back up to Washington by sea. McClellan absorbed the blame for his apparent irresolution with his lips tight-clenched. He was a competent commander who well knew that he could have ended the war at that point; but he was forced to follow orders from a higher command; PAGAN. After this, the bloody struggle dragged on for three more years.

The master-controller in the North, directing the war, was Edward Stanton Lincoln's Secretary of War. In the South it was probably the man who was later to become the Grand Master of the Southern Jurisdiction of Freemasonry, General Albert Pike. While more than 2,000,0000 Christian soldiers on both sides were driven to wound and kill each other, these two satanic monsters prolonged this war of mutual annihilation till the South lay prostrate.

The first goal of these fiends was to crush the power of the fundamental Protestant Christian Churches which were then flourishing in "The Bible Belt" of the South.

Second, it was to massacre as many Protestant Christians as possible on both sides and cripple Christian Culture.

Third, it was to undermine the people's fundamental belief in divine providence and to sow the seeds of doubt; beginning the process of converting a fervently Protestant Christian nation into toleration of the public expressions of atheism we see today.

Fourth, it was to enforce the assertion that there exists a legal bar of separation between church and state. This, in practice, meant that Protestant Christians are now barred from any real role in government

Fifth, and finally the war was fomented to eradicate the principle of States Rights; thus giving supreme power to a centralized Federal Government.

Thus, ended American Republic.

THE ROAD TO EMPIRE

Following the Social War (88 BC), while the Roman Senate attempted to cope with its internal problems, population growth and of administrating the widely dispersed provinces represented an increasingly difficult task for the old Republican system. As each new crisis arose, a succession of military leaders intervened to reestablish order. Each one attempted reforms but they could not resolve the fundamental organizational problems. For a period of thirteen years, (40-27 B.C.), Pompey, Octavian, and Marc Anthony waged a complex civil war for supremacy. In the event, Octavian, the nephew of Julius Caesar, (later called Augustus) defeated the last of his rivals.

Victory left him with the responsibility of ruling the empire as a dictator and he was sufficient to the task. He, as well as many other Romans held a sentimental attachment to the ideals of the Republic but they determined that the establishment of an efficient and stable administrative system was of primary importance. The progress toward absolutism was gradual during Augustus' reign. At each juncture administrative changes were implemented, as a temporary measure, to answer to some emergency. Yet when the emergency passed it was found impractical to repeal these laws. It was Augustus' goal that the Republic be reestablished but conditions made this impossible. Rome's territories now extended from the English Channel to the Euphrates while the semi-civilized barbarian hordes pressed in continually upon the Northern frontiers. Rome was seldom at peace.

An important psychological factor in every political dictatorship is the "Cult of Personality." It is a fundamental weakness of human psychology that it wishes to identify an organization or a government with the person of its leader. As democracy deteriorates into mobocracy the people become sickened by the excesses of freedom. They begin to wish, in their hearts, for "order" and for a man of authority who will give them order. Eventually, it comes about that "Free Men" finally cry out for Caesar; (In our times, for a Stalin, a Hitler or a Big Brother). Augustus, in the eyes of his people now symbolized Rome. Had he not provided peace and order? Should he not be worshiped as a god? Unable to deny the political value of this popular belief, the successors of Augustus had little choice but to continue to govern by assuming the title of Caesar and by asserting that they ruled by their own divine right.

As each succeeding Emperor expanded the reach of the executive branch the Senate was reduced to a rubber stamp. Dutifully, it ratified whatever laws Caesar and his administration wished to enact. This created "a revolution within the form." While public support required the fiction of a representative republic, it was no more. It had been replaced by a socialist dictatorship.

AMERICAN PARALLELS

Abraham Lincoln might accurately be styled America's Julius Caesar. Both men were political experts driven in part by noble ideals; ideals which in both instances led to their deaths by assassination. Whatever Lincoln's commitments were to PAGAN, before he was elected, during the War he was influenced by its agent the Secretary of War Edward M. Stanton; the man who eventually became his Brutus.

New evidence indicates that the conspiracy to kill the President and his Secretary of State William T. Seward was ordered by Pope Pius IX from the Vatican in Rome. Using their Jesuits in Montreal, they employed John Wilkes Booth and his band as their assassins. At the same time as the President lay dying a second team of murderers, led by Booth's confederate, Louis Powell, broke into Secretary of State Seward's home and severely wounded both him and his sons.

According to a recent history ("Blood on The Moon" by Edward Steers Jr. University Press of Kentucky 2001) Booth's plan was said to have been at first to only kidnap Lincoln and take him to Richmond to be held as a hostage; to gain better terms for ending the hostilities. At least that is what he told his fellow conspirators. But, if that were true, one must ask why then was the plot continued after General Lee had surrendered and the war was nearly over? What political purpose would it serve? Was the plot continued only because of Booth's desire for personal revenge?

The controller of the plot, we now know, was the infamous Dr. Samuel Mudd who sheltered Booth on the night of the murder and who set Booth's broken leg. Mudd's descendants, to this day, insist that he was innocent of any conspiracy; but the true story appears to be that he had twice met with Booth before the assassination; making arrangements to assist him to flee afterward. Dr. Mudd *was* a Southern sympathizer; but more importantly he, John Willkes Booth, Mary Surratt and her son John Surratt, and all the other conspirators were Catholics. Booth and Mudd first met and conferred after they attended Mass together at St Mary's Catholic Church in Bryantown, Maryland in December 1864. (Blood On The Moon. p. 74.) At that time Maryland was the most Catholic State in the Union having been founded by the Catholic Lord Baltimore in the 17th Century. (Booth's original home had been in the City of Baltimore.) And, if Booth was so fervently a Southern loyalist, why had he not enlisted and fought for the Confederacy?

While the author of "Blood on the Moon" is painstakingly thorough in reciting every name place and date in the assassination account, he leaves the final conclusion to his readers who must discover for themselves the religious motivation. The facts indicate that the plot was authored by the Papacy and the Jesuits by using Booth and his friends as willing tools to carry out the removal of Lincoln. With this approach, Booth's actions begin to make sense. We know it is the policy of the Jesuits to kill any political leader who stands in their way. (i.e. later, Presidents Garfield, McKinley and the two Kennedy brothers) The Jesuits are taught that murdering the enemies of the Pope is virtuous. Lincoln's desire to act in a Christian manner toward the defeated South by permitting those States to soon resume their former role as States within the Federal Union was in opposition to the Papal/Masonic plan. Their plan was to reduce the Southern States to vassalage. John Wilkes Booth was so brain-washed by the Jesuits that he believed that by killing Lincoln he would be seen to be an heroic figure. Booth fully expected to be helped to escape. But once he was no longer needed, he and the other conspirators were left to make their own ways. All but one were rounded up; tried and found guilty. Four of them were executed forthwith to silence them permanently.

The religious motivation explains why Secretary Seward was also made a target. Seward was a Protestant Christian and well aware of the evil forces about him. While a young legislator in the State of New York he had been a member of the Anti-Masonic Party. As Lincoln's Secretary of State he served his country well and was Lincoln's most trusted advisor. Also, as Secretary of State, he was in the line of succession to the Presidency. Such men as he and Lincoln were not wanted in the government and were to be eliminated.

To understand who ultimately pulled the strings one should ask who benefited? The ones who benefited were not John Wilkes Booth and his simple-minded accomplices but rather it was the Secretary of War Stanton, the Masons and the Pope who, by removing Lincoln, took one more step down the road toward their One World *dis* Order. Their purpose was to fuse the separate States into one Federal Government that could, thereafter, be infiltrated and controlled by their agents. Lincoln and men like Secretary Seward would have blocked that plan.

The one conspirator not caught was young John Surratt, Mary Surratt's son. Before the war he had been in a Jesuit seminary studying to be a priest. His role in the conspiracy was to act as a courier for the group. His repeated trips were to bring back money and orders from the Jesuits in Montreal. When the assassination took place, he fled back toward Quebec (i.e. Catholic Canada) where he knew he might find safety. Realizing he was wanted by the United States government, with a price on his head, he sought refuge in a Catholic church in the little town of Liboire, Quebec. There, he was hidden for several months by a Father Charles Boucher. Then, in August 1865, he was taken to the City of Quebec by another priest, Father LaPierre, who put him on board the ship, "Peruvian," a vessel that carried him on to Liverpool, England. There, once again, he found shelter with papal agents. Continuing on further, with, what we must presume was Papal assistance, he arrived, finally, in Rome where he was given shelter within the walls of the Vatican itself! To keep him occupied he was enrolled into the Pope's personal military guard, the Zouaves! (See photo Ibid: p. 231) What more convincing proof does one need to understand that the whole Booth/Mudd assassination conspiracy was, in reality, a plot directed and funded by the Jesuits; at the direction of that most wicked of Popes, Pius IX? This is the same mortal man who five years later had the supreme audacity to have himself declared "Infallible!"

Unfortunately for Surratt, one of his fellow Zouaves recognized him and to gain the $15,000.00 reward then being offered by the U. S. Government he reported Surratt to the U. S. Consul in Rome. At first the Vatican refused to extradite him but finally under U. S. diplomatic pressure, they did render him up. He was taken back to the United States to stand trial. However, because the public's passions by that time had cooled, the cause of justice was not well served. At his trial, a civil jury failed to convict him. Thus, this obviously guilty assassin escaped from the hangman's noose he so fully deserved.

Putting this newly revealed information into perspective and knowing that Booth from the beginning was furnished with $3,000.00 or more by mysterious (probably Jesuit) agents in Montreal from as early as the summer of 1864 and of the fact that the conspirators were all Catholic and that the Catholic Hierarchy of Canada and of the Vatican itself sheltered the one remaining conspirator who could have revealed their criminal involvement, is it not clear that the assassination of President Lincoln was not the work of one insane man nor of the Confederate Secret Service which, because the war had ended was being disbanded, but rather it was a Papal assassination; a "final decree" that cold-bloodedly murdered President Lincoln because he would have thwarted their plans for the United States of the future? Lincoln was never a church-going Christian but he was beginning to think and act like one. For that he was executed.

Was Stanton involved? Certainly he was by his acts of omission. At the end of the Civil War his Secret Service police force headed by the detective Pinkerton was fully aware the plot was underway. Stanton for his own reasons allowed it to continue. That the Pinkerton's knew of the plot and the of identity of the plotters is shown by the fact that within a few days after the assassination they had rounded up and arrested all of the conspirators except the one who had escaped to Canada.

During the ensuing inquiry and trial, Lincoln's and Seward's attackers and all the plotters were tried in secret by a military court. At all times both in their cells and during their trials and executions their heads were covered with leathern hoods that prevented any possibility of them speaking to their guards and revealing the identity of their employer. By these means the "politically correct" version of the Lincoln assassination became the official version.

Lincoln was killed by PAGAN because he did not intend to follow their direction. His policy toward the defeated South was to be a program of moderation; permitting the Southern states to retain their prior rights; all except the right to secede. His plan was to bring the Southern States back into the Union as soon as possible; permitting them immediate representation in Congress; and allowing them to accomplish resettlement of the freedmen at the local level. Lincoln had indicated he would follow a policy of reconciliation with the defeated Southern States.

On February 2 1865 while the siege of Petersburg continued, the President General Grant, and Secretary of State Seward met with a Confederate delegation on the President's yacht in Hampton Roads on the James River to discuss terms for peace. At this meeting Lincoln said that if they ceased the rebellion the seceding states could rejoin the Union and the liberation of the slaves could be accomplished at the state level over a period of five years. He further told them certain unnamed persons had offered to put up $400000000.00 to compensate Southern slave owners for their financial losses. But although the position of the South was already hopeless with General Lee's Army boxed in at the siege of Petersburg, the Confederate delegates refused these reasonable terms and said they would yield only upon the condition there be two separate nations not one Union; a condition unacceptable to Lincoln. This unreasonable demand cost the South dearly over the next twenty years. (Carl Sandburg. "Lincoln." reprinted in the "Lincoln Reader" Rutgers University Press 1947 p. 500.)

Lincoln had presented his plan for reconciliation in his Second Inaugural Address delivered just a month before his death.

> "With malice toward none; with charity for all; with firmness in the right as God gives us to see the right let us strive on to finish the work we are in; to bind up the nations' wounds; to care for him who shall have borne the battle and for his widow and his orphan—to do all which may achieve and cherish a just and lasting peace among ourselves and with all nations." (Ibid. p. 493.)

But this was not the intent of PAGAN. Their plan, endorsed by the Northern Radicals, the Jesuits, the Abolitionists, the Liberal Churches, the Unitarians, the Congregationalists the Union League Club and by their master-agent Secretary of War Stanton, planned a draggonade of the Christian Southern States by treating them as a conquered foreign enemy. Lincoln's death freed them from the restraints of the Executive Branch. Stanton and the Senate Radicals set up a program whereby Federal troops would occupy the South to protect and assist the agents of the Freedman's Bureau (the carpetbaggers) in carrying out the strict mandates of the Emancipation Proclamation. White citizens were disenfranchised while their former slaves were immediately given the vote. Until they met all the requirements set forth by PAGAN the Southern States were not permitted to reenter the Union nor have representation in Congress.

During the next four years, Stanton and the Senate ruled supreme. President Andrew Johnson, who favored Lincoln's approach, was virtually stripped of his executive powers by Stanton and the Senate Radicals. The Reconstruction laws and the Freedman's Bureau Act were both passed over his veto. When Johnson attempted to dismiss Secretary Stanton he was told he did not have the power to do so without the consent of the Senate. Stanton refused to be dismissed and for the rest of Johnson's term he was virtually dictator of the United States; going so far as to attempt to remove President Johnson by impeachment.

THE BEGINNING OF EMPIRE

As with Rome following the death of Caesar, the U. S. Congress seemed to have retained all of its powers but behind the scenes a hidden force had become the true authority. From the Civil War until World War I, two broad lines of popular thought contended with each other. The large majority of the American people were traditionally religious, moral and peace loving. Theirs was the fundamental belief that the American Nation was unique; that it was a land blessed by God; given the fruits of justice and liberty; made available to all. With this they were satisfied. America had created the perfect form of government and should be left to develop itself to the full within the land God had given it; and it should ignore and repulse the political diseases of the outside world. The governments of Europe and Asia were wicked and should be kept from our shores. Foreigners were not to be our exemplars; rather we would be theirs. This was the popular myth.

Opposing these naive convictions were the secret plans of a small but influential minority: PAGAN and its "Grand Design." Its plan was to expand the reach of the American Empire until it would, eventually, dominate the world. We see the implementation of this policy in the late 19th Century with the re-affirmation of the Monroe Doctrine that asserted that all nations in the Western Hemisphere must be American protectorates. The idea was also stated that America had a "Manifest Destiny" first to conquer the Western Frontier, then the Pacific Ocean and eventually to bring the benefits of American Culture to the benighted nations of Asia.

By the conclusion of the Spanish-American War, again, a war contrived for the purpose, the last of these goals was realized. America annexed Hawaii; the Philippines were seized from Spain and U. S. Marines were on station in Peking. By a similar "tour de force," Theodore Roosevelt, by using his "big stick' and by clandestine means, authored an artificial revolution in the nation of Columbia to create a new nation, the state of Panama; whose puppet leadership immediately signed a treaty giving the United States the right to construct a Canal across the Isthmus. Finally, the U. S. would have its "two-ocean navy." A popular slogan at the time was: "Westward The Course of Empire."

Time scale: (Rome 27 BC=USA 1904 AD

After these easy conquests, as in the case of Rome at the beginning of Augustus' reign, it became more convenient to transfer effective power into the executive branch. 1913 (the beginning of Woodrow Wilson's presidency) formed the watershed year between the old and the new. That year saw the centralization of finance and money power under the foreign-owned Federal Reserve System. It, also, saw the founding of the Internal Revenue Service by means of the 16th Amendment and, more significantly, it saw the passage of the 17th Amendment as well.

Few now recall that under the original Constitution the function of the Senate was to act as a mature and stabilizing body to restrain the tendency toward political imprudence that might occur in the popularly elected House of Representatives. For this reason the Constitution ordained that the Senators were not to be elected by popular vote but that they were to be appointed by the Legislatures of the several States. The effect of this provision was that Senators appointed by their States could act as a check upon the excesses of the House of Representatives and upon the Executive and Judicial Branches. The Senator from Ohio spoke for Ohio; The Senators from New York *were* New York in the U. S. Senate.

The Senate was the dominant body. By refusing to pass legislation or to ratify treaties or cabinet appointments it could check populist tendencies that might influence the decisions of the House of Representatives and the Executive Branch.

The passage of the 17th amendment requiring the direct election of Senators by popular vote, at-large, in each state was a master-stroke by PAGAN. It undercut the dominant position of the Senate and reduced it to the status of a senior division of the House of Representatives who, like them, now served at the will of PAGAN. Because the Senators now face the enormous expense of being popularly elected and reelected, state-wide, they are made or broken by the financial support given or withheld by PAGAN. The U. S. Senate is now a hollow shell; a mere sounding board for the hearing of citizen complaints and a forum for the

Senator-ombudsmen to explain the latest regulations of PAGAN to the general populace.

As in Rome, when the Senate, that pillar of the old Republic was gutted, many applauded this advance in popular "democracy;" not recognizing that by permitting this "revolution within the form" they had delivered themselves into the hands of Caesar.

Time scale: (Rome–14 BC = United States 1913).

THE GREAT WAR—1914–1918

Following the warning shocks given them by the Napoleonic wars, and for the next one hundred years, the nominally Christian kings of Europe carried on a rear-guard effort to maintain their traditional political systems. But, while these monarchs believed they *"ruled"* their countries, the ground was being mined beneath their feet. Dreaming of national honor and military glory, they led their subjects onto the killing fields prepared by the manipulations of PAGAN. When one considers the meaning of the sacrifice of all those millions of persons who suffered and died in "The Great War" what do we see? We see its military cemeteries filled for the most part with crosses marking Christian graves.

> "In Flanders' fields the poppies blow
> Between the crosses row on row
> In Flanders' fields—"
> (Colonel John McCrae, d. 1917)

"The Great War" was a triumph for PAGAN. They were able to cause the death of perhaps 20,000,000 Christian soldiers and civilians in a massive blood-sacrifice to their god—Satan. This war pulverized European society and caused such deep disillusionment it prepared the soil for new governmental systems based, not upon Christian morality, but upon the new religion; the religion of the godless Nation State.

What was the purpose of the war? Ask what was the result? The result was that Christian monarchy and all it culturally represented in Europe was eradicated. PAGAN now dominated the newly created and balkanized "democratic" nation-states they had organized upon ethnic lines, out of the ruins. The grand monarchs of Germany, Austria-Hungary, and Russia were, one and all, either dethroned or executed. PAGAN then set up the nominal nation states of Poland, Czechoslovakia, Hungary and Yugoslavia as a sop to those peoples' unrealistic dreams of independence; when any understanding of the principles of realpolitik would show to them that, as nation-states in the twentieth century, they were only to be "temporary" pawns in the game. The destruction of monarchy permitted PAGAN to take its next step toward its ultimate goal by establishing a "League of Nations" and afterward the "United Nations." These final steps abolished nationhood.

The end of the Great War in 1918 is a corollary to the defeat of Marc Anthony and Cleopatra by Octavian (Augustus) at the battle of Actium in 31 B.C. Those dates marked an end to the power of both the Roman, modern European and American religion-based governments. They declared the transformation of these states into departments of a new, One-World *Dis*-Order.

Time Scale: (Rome 31 B.C = Western Culture 1918)

CHAPTER SEVEN

EMPIRE

Following the Augustan Revolution, the Senate of Rome seemed outwardly unchanged but this concealed the fact that an Imperial administration had come into power. In the United States a parallel metamorphosis began in 1913 with the founding of the foreign-owned Federal Reserve Bank and with the establishment of a federal income tax and with the emasculation of the U. S. Senate. These changes inaugurated a new system of government. As in the Augustan Revolution it was a revolution "within the form."

Similarly the American public could not imagine that the Constitution had been altered yet it became apparent that new methods were being followed. Soon a "mystery man," Colonel Edward M. House, took up a position by the side of the wooly-minded Wilson as his confidante and advisor. Since then, other "mystery men" have continued to be found at the side of each succeeding President.

The public wondered; who *was* this unelected figure that "advised" the President? Why did the President need advice? House soon became, de facto, the Secretary of State overshadowing Elihu Root who nominally held the post. In 1915 Colonel House traveled to England as Wilson's private emissary to consult with Lord Grey, the British Foreign Secretary. They met to plot the steps by which America would be brought into the War. At that time, Americans were neutral and wanted no part in Europe's wars. To PAGAN, the American people's desires were of no consequence. PAGAN had determined that America must be brought into the war so they could use the power of the United States to dictate the peace.

THE LUSITANIA INCIDENT

Because the political regime in the United States was not yet as absolute as that of Great Britain, wherein PAGAN could declare wars, unilaterally, by employing the "Sovereignty" of the Crown; the American system required that the public be persuaded to fight by the creation of some atrocious incident to excite its military passions. To Machiavellians, if an atrocity does not present itself, it needs to be invented. The necessary atrocity was supplied by the sinking of the British passenger liner, Lusitania, in 1915. That sinking was arranged by a conspiracy between Lord Grey, British Foreign, Secretary, Sir Winston Churchill, First Sea Lord, and Col. Edward M. House, the U.S. President's personal representative. It was effected in this manner. Under normal practice the speed of a liner such as the Lusitania would enable her to out-run the submarines of that day, and by staying far out to sea, to be reasonably safe. On this voyage, however, under orders from the British Navy, the liner was forced to make its course close in to the Southern shores of Ireland where, by pre-arrangement with the German Navy or by the publication of this unusual course, a German submarine lay waiting. The rest is known. The liner was torpedoed and it sank with a great loss of life.

This hideous act of human-sacrifice succeeded wonderfully in affecting public opinion in the United States. The previously neutral Americans were now emotionally forced to choose a side in the war and, thereafter, they favored the Allies.

During the early years of the war the Zionist Jews in America and England were sympathetic to the cause of Germany because Germany was fighting their mutual enemy; the Russian Czar. To alter their sympathies, the British Government in 1916 issued the Balfour Declaration promising the land of Palestine for a future Jewish homeland. For this, the Zionists changed sides. Thereafter they supported the Allies.

With these preparations accomplished, the way lay open. Although Wilson was reelected, in 1916, upon the slogan that "he kept us out of war;" 30 days after his second inauguration, he went before Congress and asked for a declaration of war against Germany and the Central Powers.

When the American Army arrived in France and went up to the front line in large numbers, they were ordered into almost suicidal attacks against the German's prepared positions. The result was a horrendous slaughter. In the 10 months of 1918, during which the Americans were involved, they lost 150,000 dead! This ghastly human sacrifice was, in the eyes of Wilson, necessary to give America the right to participate at the peace table.

Through Wilson, PAGAN announced a Fourteen Point Plan for Peace which included as its major goal the establishment of a League of Nations; a supranational government. After the War, the President and the left-wing intellectuals propagandized the country in favor of joining the League but the American citizenry, and their only partially democratized U.S. Senate, recoiled in horror at the prospect. The people and the Senate still believed in the moral superiority of the American system over the corruption of European secret diplomacy. Membership in the League was rejected—but the seed had been planted; (Two steps forward, one step backward) and after a further psychological re-conditioning, effected by the 1930s economic depression and by another World War, resistance was weakened and in 1946 the American people sullenly accepted membership in the League's successor, the United Nations.

THE 1920s

Following the Great War, the U.S. entered into probably the most important decade of its economic history. Aided by a gold-backed currency and by the improvements in technology, industry and commerce prospered as never before. This was the decade that saw the construction of perhaps one-half of the major skyscrapers now standing in our cities. It was the decade during which almost every middle or working class family first came to own an automobile and the decade when the national highway system was built for these cars to travel on. With little income tax to pay, successful middle-class families suddenly were able to build and live in large homes and mansions—imitating the manners of the European aristocracy. Every man it seemed could be a king! The evidence is there. The luxurious suburbs of our major cities were first created during these years: Westchester N.Y.; Euclid Heights Ohio; Grosse Pointe Michigan and Winettka, Illinois, etc. to house these "nouveau riches."

By 1929, America had reached the very summit of free-enterprise capitalism. Those who lived through that decade recalled it wistfully as "The Golden Age." Satisfaction with the status quo was universal—except for the purposes of PAGAN. They determined there was too much economic freedom and that the citizens must be taught a lesson and brought under control. Plans for a "New World Order" were in preparation that could only be put into practice by the application of a sufficient amount of pain.

That was the reason for "The Great Depression." In 1929 the stock market was deliberately sabotaged. The "Crash" and the ensuing years of economic depression were wholly synthetic events; diabolically created for a political purpose. That purpose was to prepare the soil for the New Deal revolution to follow.

While it is true that, in 1929, prices on the stock market were optimistic, the optimism was justified. America's economy was rapidly expanding and the future seemed bright. The only technical weakness was the dangerously low level of margin requirements (10%) which encouraged speculation. This weakness could have been corrected by a judicious and gradual increase in margin requirements. That simple remedy would have restrained speculation without a disastrous effect upon prices. But the goal of PAGAN was otherwise. They seized upon this weakness as their point of attack.

Many, such as Joseph P. Kennedy, a prominent Catholic insider, were informed in advance that the Crash was being planned. The liquidity of the Stock Market then and now is dependant upon a steady supply of "call money" daily credit supplied by banks used to fund daily trading settlements and margin account debt. These funds bear interest; usually at prime rate or somewhat less. Without liquidity, the markets cannot function. It was the withdrawal of all call money funds from the brokerage houses on October 27, 1929 on the orders of PAGAN that precipitated the stock market crash. Without liquidity, those who owned stock on margin were called and told they must immediately bring in 100% cash or be sold-out "at the market." Since funding was unavailable, everyone on margin was forced to sell at once into these falling prices. A panic ensued. After 1929, prices continued to fall until in July, 1932, the Dow Jones Average hit bottom at 40. As a corollary to the market panic, commercial banks either called in their loans or refused to renew them. Without liquidity, even large firms were forced to close and put their employees out on the streets.

With his advance knowledge, Joseph P. Kennedy made a vast fortune selling short. That fortune is now held in a wholly tax-exempt (now illegal) "dynasty trust" that to this day continues to support his numerous descendants who, without the necessity of earning their livings or paying taxes, are able to engage freely in political activity. In 1934, Joe Kennedy was appointed by Roosevelt to be the first Chairman of the Securities and Exchange Commission. Some said "this was like putting a fox in charge of the hen-house."

During the massive waves of unemployment that followed, small businesses were forced to close because their customers could not pay their bills. The same inability to pay cascaded through the whole economy and was accompanied by wholesale bank foreclosures upon every sort of real property including the homes of the unemployed. When even that collateral proved worthless, the banks themselves collapsed in turn.

In his autobiography, "It Took Nine Tailors," the film actor, Adolph Monjou, wrote of his experience with the 1929 crash. He had begun his career in silent films and, by 1929, had accumulated a sizable amount of capital. Like everyone else, he was invested in the stock market. During the summer of 1929 he vacationed in France. While there, as a successful native son, he was entertained by many French notables. Chatting one night at a reception with a Parisian banker, he was asked if he was aware that the New York Stock Market was to be crashed in the coming fall? He was told that if he were invested he should exit his positions immediately. Menjou was astounded but heeded the advice. He cabled his brother in New York, who handled his portfolio, telling him to sell everything. Reluctantly, the brother complied; converting the cash into gold coins; the French peasants' traditional refuge in uncertain times. Menjou continued to be a successful actor but never again did he trust the stock market or the banks. Throughout the Depression he kept his hoard of gold in the only place he trusted; in the basement of his Hollywood home. (Menjou Adolphe "It Took Nine Tailors")

After three years of economic paralysis the normally conservative, but now desperate, American people were willing to accept anyone who promised relief.

Enter the American "Augustus," Franklin D. Roosevelt, and his "New Deal." Roosevelt, a wheel chair bound cripple, (a fact kept totally secret from ordinary Americans by the media during his lifetime) was the willing servant of PAGAN. For the vainglory of being titled "President of the United States," Roosevelt accepted direction by the agents of PAGAN. These agents included his wife, Eleanor Roosevelt, and a series of "mystery men;" the first of which was the left-wing "journalist" James Howe. After Howe's death, he was replaced by Harry Hopkins. Hopkins together with other PAGAN agents such a Felix Frankfurter from Harvard University, Harold Ickes, Henry Morganthau Jr. and Henry Wallace formed the so-called "Brain Trust" about Roosevelt; to help him implement the revolutionary agenda.

During the first "One Hundred Days" of the Roosevelt administration, in the spring of 1933, Congress passed into law all of the fundamental legislation required to convert the United States into a Socialist Empire. In truth it *was* a "New Deal;" except that the dealer was dealing from the bottom of the deck.

In the name of this contrived emergency, hundreds of agencies were set up designed to control every conceivable economic and social relationship. The country was taken off the gold standard—permitting the Federal Reserve to print counterfeit money at-will. Huge administrative bureaus were created; most of which survive to this day. The AAA, the NRA, the SEC, the NLRB, the WPA, the FHA, the FDIC, and the CCC etc. etc. Each purported to have a beneficial purpose but they, in fact, overrode the rights of the states and of the citizens by federalizing nearly every aspect of social and economic intercourse. Thus was consummated the seventy-year process of centralization extending from the Civil War to 1933.

Since 1933, little has changed. American politics and governmental policies, since, have only been fine-tuned under a totalitarian administration to ever more completely circumscribe American society's every thought and action.

The Republic is dead; legislated out of existence during the "Hundred Days." The two elected bodies, the U. S. House of Representatives and the U.S. Senate comprise only 535 easily controlled individuals who, no matter how well intentioned, are mere relics of the past. They have no real power over the 7,000,000 persons on the Federal Executive payroll who administrate and *are* the U. S. Government.

If these "legislators" have any function at all it is as ombudsmen; reviewing the latest administrative refinements as they are enacted into law and by inducing their constituents to accept them. The Administration writes the laws and the Congress applies its rubber stamp of legality.

Time scale (Rome AD 14= USA, AD 1933)

In 1945 at the end of the Second World War with the instrumentality of the atomic bomb in its hands PAGAN effectively completed the military conquest of the world. History, since then, has been the consolidation of this conquest. The endless series of local conflicts, revolutions and civil wars that appear to the public as insoluble problems; are wholly created and funded by PAGAN through its secret army, the CIA, for its own unique purpose. That purpose is to maintain a psychological state known as the "Red Terror;" a plan originally devised by the wizards of Jesuitical-Zionist-Marxism in Russia.

The "Red Terror" is a program to make war not upon foreign enemies but upon the innocents of one's own country. Forced to live under a permanent state of terror, the people will more willingly accept the eradication of their civil rights. This policy is in effect today. The violence, the officially sanctioned attacks upon morality and "terrorist atrocities" are all part of the "Red Terror." It is the programmed massacre of the innocents.

Observe how, in every conflict, whether be it in Afghanistan, Vietnam, Beirut, Israel-Palestine, Central America, in Belfast, Africa or on the streets in the United States, all have one feature in common. Who is it that suffers and dies? It is not the political leaders or the terrorists; the truly guilty. No, it is the innocent civilians; men women and children who are raped robbed beaten or blown to pieces while simply trying to have a peaceful meal in a restaurant or pub. This state of terror continues on because PAGAN is supplying the arms, the explosives and the money to both sides. In this state of perpetual hysteria, PAGAN rules.

After an atrocity such as the destruction of the World Trade Center in New York City in 2001 who is it who are punished? Not the alleged "terrorists" who are never to be found. No! It is the American people who are being crushed under new dictatorial legislation.

In 1991, in a military charade called the Persian Gulf War, ("Desert Storm") we massacred thousands of Iraqis. This "war" in truth was another international military exercise; carried out for the sole purpose of testing new forms of weaponry on live targets. It was the Iraqi civilians and their common soldiers who were massacred by the latest "smart-bombs" and guided missiles; not their leader Saddam Hussein. Saddam, who secretly is an agent of the CIA remained unharmed. This war however made him the unifying idol of the Islamic peoples vs. Western Imperialism. (Exactly the result desired)

These synthetic wars and crises have provided the "news" during the late twentieth century. Historically, they parallel the interminable Roman wars of consolidation during the First Century AD. Rome's campaigns were waged against the

Germans, the British, the Parthians and the Jews, to eliminate every potential enemy that would not accept its rule. In the Twentieth Century the wars and created crises, from the end of First World War to the present, have had as their overriding purpose the physical reduction of all peoples beneath the will of PAGAN.

While military in nature, these wars are really waged for religious purposes; to consolidate their conquests by converting the subject population's former religious traditions into the "modernist" attitude of unbelief.

What is the result of this unbelief? What is the official state religion of the majority of nations today? Nominally it is agnosticism—neutrality toward religion. But just below the surface it can be seen that it is Jesuitism and its offspring Freemasonry, in manifold forms of concealment. From the lowest to the highest the public officers comprising the administration of government in most nations is entirely carried on by Masons or Jesuits.

A little research will prove this to you in your own locale. Should you develop an ambition to serve in public office, it will soon become apparent that to "get elected" one must be a "joiner." What does it mean "to be a joiner"? What it means is that you must first join and be a member in good standing of a lodge of the Masons, or of the Knights of Columbus or of B'nai B'rith, before the doors of political opportunity will open to you. This is a fact not appreciated by the ordinary voter. But in any political jurisdiction, careful observation will reveal that nearly every officer; from the County Coroner, the County Sheriff, the City Councilmen, the County Commissioners on up to and including all the higher City State and National offices; including the office of President of the United States are all filled by Masons, Knights of Columbus, Mormons or "Sons of the Covenant." The legal profession and the judiciary are, of course, almost entirely "Members of the Craft."

And what is the religious position of Masonry? The religious position of Masonry, nominally, is "toleration" but their toleration devolves in practice to mean moral relativism and atheism. They define theirs as "the religion upon which all men agree." (that of course is no religion.) Their oaths and loyalties to Masonry are placed ahead of any other moral commitment. Like the Jesuitical system from which it sprang, Masonry exempts its member from the normal restraints of truth and honor. Masons believe they are "The Masters of the Universe" who, through their rites of perfection, have become demi-gods; beings above good or evil.

This absence of any moral restraint explains how President Franklin D. Roosevelt, a High Mason, (McCullough, David, "Truman" New York: Simon and Schuster, 1992 p.325.) could have participated in facilitating the surprise attack upon Pearl Harbor. It explains, also, how another Mason, President Harry S. Truman, (Ibid p.78.) could order the dropping of two atomic bombs upon

Japanese civilians (just as a scientific experiment) when he knew the Japanese Government was trying to surrender. To such men, the morality of those decisions was secondary. Their orders came from PAGAN and those orders overrode all moral restraints—and they were obeyed.

PEARL HARBOR

More than sixty five years after the event, it has become known that before the Pearl Harbor raid the Americans had broken the Japanese "Purple Code" so that our highest officials knew the day and the hour of the impending attack. Nevertheless that information was withheld from the Army and Navy Chiefs in the field.

What has not been revealed before is that, upon direct orders, the U. S. Navy ships were deliberately *prepared for sinking*! One question that has never heretofore been asked is: How could these armor clad and compartmentalized U.S. battleships have been sunk so easily by one or two torpedoes? Weren't they built to absorb such damage and to yet survive? How is it that when hit they sank so rapidly that their crews had no time to even evacuate the lower decks? Recently an explanation has come to light.

A retired naval officer who, at the time of the attack, was an enlisted man reported that in 1941 the standard practice for the Pacific Fleet was to operate as two Divisions; one cruising at sea while the other stayed in port for refitting and refueling. But in late November 1941 the program was changed. Both Divisions were ordered into port with all the ships being tied up in rows at the docks. Only the aircraft carriers were at sea. The men were then told to stand down for hull inspections which meant that to make those inspections all the top cover plates sealing the double-hulled compartments in the bottom of the vessels had to be removed; thus destroying their water-tight integrity. Thereafter any major leak anywhere in the hull would become uncontrollable. That is why when hit by even one torpedo they sank like stones.

Therefore, the battle at Pearl Harbor does not represent a so great a victory for the Japanese Fleet or so great a defeat for the U.S. Navy. What really occurred on December 7, 1941 was that the Japanese carrier pilots enjoyed the privilege of launching their torpedoes against undefended and deliberately crippled targets. The United States Pacific Fleet was sacrificed along with three thousand of its sailors for the single purpose of creating an atrocity, ghastly enough, to bring the American people into the war. (Source name withheld)

The next day, December 8, 1941, President Roosevelt made his famous address to the Congress calling for a declaration of war against Japan saying that; "December 7th 1941, will be remembered as a day of infamy;" referring to the

Japanese attack; even while he knew that he and PAGAN were the very authors of that infamy. This speech must be recorded in future history as the very apogee of hypocrisy.

What followed was an unprecedented outpouring of wealth and blood. Five years were torn out of the lives of the young men and civilians of the nations on all sides who were forced to become combatants. Millions of soldiers and civilians died under hailstorms of bombs, shells and bullets for reasons they never understood; set up like ninepins and struck down.

Those who lived through those years will never forgot them. The names of the battles and campaigns form a litany of human agony. For the Americans in the Pacific there were Tarawa, New Guinea, Buna/Gona, Guadalcanal, The Slot, The Coral Sea, Midway, Bouganville, Saipan, Leyte, Okinawa and Iwo Jima. In Europe another list no less remembered: The Blitzkrieg, Dunkirk, The Battle of Britain, Tobruk, El Alamain, Tunisia, Sicily, Anzio, Normandy Beach, St. Lo, the Moselle Crossing, the Hurtgen Forest, and the Bulge.

For the soldiers, sailors and airmen of Germany, Italy, Poland, and Russia the cost was no less dear. The battles on the Eastern Front ring out their toll as well: Smolensk, Leningrad, Moscow, Kiev, Kursk, Stalingrad, Warsaw and Berlin. This Eastern war pulverized the cultured peoples of both Europe and Russia; leaving them prey to reorganization into new "democracies" under the ever-tightening grip of PAGAN.

As in the Old Roman Empire of the 1st Century AD, the conquered states, after being militarily reduced, are to be integrated into the single world government; a New Roman Imperial System. But while PAGAN'S triumph may be nearing completion, there remains within the world's population one massive hidden enemy as yet unconquered. It is the enemy of religious faith. For while the bodies, the labor and the wealth of the conquered peoples may be demanded, the control of men's minds is a more difficult undertaking.

It is on this level that PAGAN must be opposed and defeated. The Old Roman Empire ended; "not with a bang but with a whimper." It came about when the peoples of the Empire ceased to believe in it. It ended finally when the proclamations of the last Emperors were met with derision.

When *its* orders finally cease to be obeyed, the days of PAGAN will likewise be numbered. As in the 1st Century AD, there remains a counter-force represented by the bible-based Christian underground movement that like the underground church of ancient Rome will bring down this Second Roman Empire as well. PAGAN is aware of this. The battles of the future will be fought not for men's bodies but for their souls. These will be the "Wars of Religion." Indeed they have already begun. Time Scale: Rome (100 AD) = America (2000 AD.)

CHAPTER EIGHT

POLITICS AND PERSONALITIES

SPLITTING THE PARTIES

A technique often used to control the outcome of close elections is to weaken the political party to be defeated by entering multiple candidates upon that side; thus splitting that party, and ensuring the election of the single candidate in the other party; the candidate programmed to win.

Therefore, PAGAN directed one of its members, ex-president Theodore Roosevelt, to feign a personal conflict with Taft over differences of principle and to then to form his own, third, "Bull Moose Party." In the election that followed, Roosevelt drew off just enough Republican votes to elect Wilson; even though Wilson still received only a minority of the popular vote in this three-way race. Republicans accurately blamed Roosevelt for the election of Wilson but never understood that it had all been done by pre-arrangement; on orders from PAGAN. Roosevelt was even directed to repeat this performance during the campaign of 1916 and, once more, he assured the re-election of Wilson.

This technique of party splitting is taken out of the closet and reused whenever needed. In 1952 the Democratic Party was split into to two factions: the Liberals and the Dixiecrats, so as to elect the Republican Dwight Eisenhower. In 1968, the Democratic Party was again split by the candidacy of the Governor of Alabama, George Wallace, thus ensuring the election of the Republican, Richard Nixon.

A more recent example occurred during the campaign of 1992. The technique was employed once more; with its usual success. President George Bush Sr., the Republican candidate, was nearly certain to win re-election because, as in 1912, the legacy of two previous Republican administrations gave him the incumbent's advantage. Suddenly, however, there arrived upon the scene, seemingly from out of nowhere, two political "mystery men." The first was the Governor of Arkansas, Bill Clinton, a political unknown on the national scene at that time and the second was a wealthy political amateur, Mr. George Perot.

PAGAN'S plan was to elect Bill Clinton so it ordered George Perot to use his seemingly unlimited and mysteriously acquired fortune[13] to form his own,

---

13   Holy Bible, King James Translation. Genesis, I, 26, 27.

Americans United Party, supposedly, to the "right" of George Bush in order to split off conservative Republican votes.

But during Perot's 1992 primary campaign, a real emergency arose that almost upset the plan. Many voters were so opposed to the liberalism dominating both major political parties that, by the end of the primary period, the fiery speeches of Perot had so inflamed the conservative voters in both parties that PAGAN'S diversion became altogether too successful. As the date approached for the final primary in California, PAGAN'S polls showed that not only was their pseudo-candidate splitting the Republican Part, he was on his way to destroying it! The polls showed Perot would overwhelmingly **win** in California with the Democrat Bill Clinton coming in last, in third place! Had Perot been permitted to continue at that pace, he might have won the Presidency—outright!

Once again, a "command decision" was necessary. PAGAN ordered Perot to immediately drop out of the race before the primary election! At the height of his popularity, George Perot astounded his followers by resigning his candidacy; thus losing the California primary. Then, more amazingly after his followers were sufficiently bewildered, in September 1992, he resumed his run as a candidate in the General Election. But now, with a weakened effort, he produced only the pre-planned result; the splitting of the Republican Party and the election the Democratic candidate; William Jefferson Clinton.

The popular vote for the Republican Bush was 37%; for the "Conservative" candidate Perot 20%; and for the Democratic candidate, Clinton, 43%. Without the splitting of the Republican vote, caused by George Perot, George Bush Sr. would have won in a landslide; with 55% or more of the popular votes. Once again an election was "stolen"—thwarting the will of the American people.

So arrogant is PAGAN, and so gullible are the voters, that in the re-election campaign of 1996, it was possible to replay the same scenario. While the deliberate non-candidate, Senator Robert Dole, ran as the Republican, Perot in a repeat of his 1992 performance, once again diverted 19% of the vote from the Republicans—thus re-electing Bill Clinton; once again with less than a majority.

DENIGRATING THE OFFICE

After installing each new president, but in order to prevent him from becoming a true leader, PAGAN next proceeds to attack and denigrate his reputation by creating real or fabricated scandals about him, his policies, and his administrative aides. As a candidate he is given the white hat of purity to wear but, once in office, he must wear the black hat of a villain. Each succeeding president submits to this process. If a highly popular president, such as a Dwight Eisenhower, or a Ronald Reagan is personally unassailable, his aides are impugned as his surrogates. The Watergate affair of the 1970s, used to attack the Nixon presidency, was totally orchestrated by PAGAN and its operatives. The staged burglary of the Democratic National Headquarters offices in the Watergate Building (a building that is owned by the Vatican) was carried out by PAGAN'S paid thugs who were meant to be caught. There was never any information in this office worth stealing and the Republicans; with an incumbent President in office were in such a strong position, politically, there was no motive for them to do this. The reporters on the Washington Post, Woodward and Bernstein, who "broke" this case, discovered nothing on their own. They were merely "messenger boys" relying wholly upon information fed to them, step-by step, from within the administration by an unnamed source called "Deep Throat." Whenever they ran out of material, or if the public's interest seemed to lag, "Deep Throat" phoned again with a new, outrageous, "revelation" to stimulate the public's anger anew.

For years, the aides of President Bush Sr. passed through the fires of the "Iran-Contra" controversy. This "scandal" involved another set of synthetic issues concerning the CIA's transferring of money and U.S. military weapons from one of their client states to another. The whole affair was a faked political exercise but nevertheless it defamed the Bush Administration and provided the media with headlines to deceive the public.

The exposés during President Bill Clinton's presidency can only be described as bizarre. Given his numerous misadventures, PAGAN was provided with enough material to defame a dozen Presidents.

PAGAN elects, and then attacks its own presidents, because its greatest fear is the "cult of personality"—that a President might actually become so popular he could break ranks and begin to act upon his own. (i.e. John F. Kennedy?) PAGAN'S goal is not to outwardly support its puppets but, rather, to maintain a state of mind-control over the American public by employing this psychological device. Every president since 1920 has submitted to this process. It is the price they pay for personal glory and a lifetime retirement plan.

These attacks are designed to brain-wash the public into believing our constitutional system of government is fundamentally flawed; that it is inadequate for this modern age and needs to be replaced! Replaced by what? Replaced by a godless form of "authoritarian" government; organized around an autocratic "brain-trust" a Central Committee that will direct a series of "councils" (i.e. soviets) extending downward to the local level which, together, will control all aspects of our economic and social lives; a Socialist Soviet styled feudal dictatorship!

Progress toward this goal continued under the Clinton administration. Whatever Mr. Clinton's shortcomings he was an effective errand boy for PAGAN. Disregarding the state of intellectual confusion maintained by the media's smoke-and-mirrors games, President Clinton accomplished many of his assigned goals.

The passage of The North American Free Trade Act (NAFTA) which merged The United States into the "One World" economic system and the passage of the law granting China "most favored nation" trading status. These acts virtually abolished the idea that the United States has any national or economic borders.

The creation of the first half of a nationalized health-care program by virtually forcing the senior-citizen Medicare system to pay benefits only to doctors and hospitals who join the government's mandated HMO programs.

His repeated attempts, at every opportunity, to pass federal gun registration laws in order to eventually disarm the citizenry

His publicly proclaiming that: "We need a New Form of Government" prepared the ground for his successor.

By his personal conduct and by his refusals to be truthful, he did much to destroy any belief in the office itself. This prepared the public for the election of the next president, George W. Bush, the man of "morality" (or perhaps the "man of steel)" (Big Brother?)

While the public's attention was focused upon the personal deficiencies of Bill Clinton he, and now his wife, Hillary, along with PAGAN continue to forge the chains of our enslavement.

WHO IS BILL CLINTON—REALLY?.

Internal evidence points to the conclusion that William Jefferson Clinton is the son of Mr. Joseph P. Kennedy, born on the wrong side of the blanket. Could this astounding theory be true? Consider the history of the man Joe Kennedy. An excellent biography of Mr. Kennedy recounts it was his habit to have physical relations with as many women as possible, other than his wife, on a regular basis. For years Mr. Kennedy traveled widely and took with him a male associate whose sole duty was to provide him with feminine companionship wherever they went. This well-known proclivity was inherited by his sons Jack, Bobbie, and Ted Kennedy and by their progeny; whose escapades are well documented.

Presumably in late 1945 Joe Kennedy visited Hot Springs, Arkansas that was, at that time, a wide-open gambling resort. There he met a young nurse, Mrs. Virginia Blyth, whose husband was a soldier still overseas. Her habit, after working hours, was to visit the local nightclubs and generally have a good time. There she met and became friends with a wealthy gentleman who just happened to be Joe Kennedy. This friendship resulted in the girl coming "into the family way." Her son, named at that time, William Jefferson Blyth, was born on August 19, 1946.

Any honest appraisal of Joseph P. Kennedy reveals that he was a rapacious businessman who seldom let moral scruples stand in his way but, nevertheless, he had one virtue. After making his fortune, his sole passion, thereafter, was to advance his sons' political careers. His overriding goal was to insert one or more of these sons into the Presidency of the United States. In pursuit of this goal, he succeeded by masterminding his son Jack's political campaigns and his rise to the Presidency. The same success might have been repeated if Bobby Kennedy had not been terminated by PAGAN for insubordination. In the case of young Bill Clinton, even though he was not acknowledged, his father had the same paternal ambitions and made full provision for him to be supported and assisted so as to achieve whatever he could; so long as he followed the guidelines set down.

Nominally, Bill Clinton is the son of a Mr. William Blyth. But Mr. Blyth, a young traveling salesman, was often away from home. It is believed that Bill was born only seven and one-half months after Blyth returned from service overseas.[14] Possibly Blyth was aware of his wife's pregnancy and suspected he might not be the child's father. This was a problem soon remedied. While driving home from Chicago, it is reported he slid off the road in a rainstorm and rolled over his Buick sedan into a farmer's muddy field. The results of this accident were peculiar in that he was not found in the car but rather his body was found face down a good distance away, in a water-filled ditch, drowned.[15] The body showed no physical injuries except a bump to the back of the head. Could foul play have been involved? Whatever—the death of Mr. Blyth on May 17 1946 blocked any further questions concerning the child's paternity.

Virginia Blyth next married a Mr. Roger Clinton who gave the boy his name. Mr. Clinton was the owner of a small automobile dealership in Hope and moderately well off. But whatever their financial circumstances there was never a lack of money for young Bill. When he was ready for elementary school he was sent, not to the public school, but to a Catholic Parochial school. This meant the payment of tuition.

As early as high school, Bill seemed to know he had a mission. While he played the saxophone in the school orchestra (reportedly he is a good musician) he also joined every organization he could; already networking for his future career. He joined the Masons as a member of De Molay.[16] This degree of political planning is most unusual in a teenager unless he was following the direction of a mature advisor. (Carpozzi George "Clinton Confidential." Emery Dalton Books. New York. 1995.)

---

14   Encycl
15   Enc
16   Author's name withheld.

Clinton next decided (or was advised) to apply to Georgetown University in Washington, D.C. because it was considered to be the best college to prepare him for a career in international affairs. (1964) Georgetown, as many do not know, is a private Catholic University operated by the Jesuit Order. Significantly, it is also the home of the U. S. "Foreign Service School" that trains our diplomats! For young Bill there was apparently no problem about the expense that must have been considerable. Tuition, room and board, at that time, probably cost $20,000.00 per year. Today, it would cost $30,000.00 to $50,000.00 per year. Who paid these bills? Seemingly, his family did not know or did not wish to know. In her autobiography, his mother, since remarried, as Virginia Kelly, stated that Bill just went off to college—period. Bill did well on both the academic and the social fronts making friends and impressing the faculty. One of his mentors was the famous Professor Carroll Quigley (of course a Jesuit) who is the author of the book "Tragedy and Hope.[17] This is the book that claims to be "a history of our times." It records PAGAN'S step-by-step progress toward world dominion. Boasting of PAGAN'S achievements, Quigley argues that secrecy concerning these matters, at least among the elite who might read his book, is no longer necessary because the triumph of PAGAN is already certain.

While still a freshman at Georgetown, Clinton, mysteriously, was called by Mr. Lee Williams, an official in Arkansas' Senator Fulbright's Washington office, and offered a summer internship. Why? Bill welcomed this as an opportunity to learn more about practical politics. His time in this job provided him with valuable contacts. Later it was a letter from this same Senator Fulbright's office to Clinton's draft board that saved him from, finally, being drafted during the Vietnam Era.

After a two-year vacation in England as a Rhodes Scholar, where he once organized a protest march "against the Viet Nam war" in front of the American Embassy in London, he returned to the United States to enter Yale University on another "scholarship;" planning to earn a law degree. Who arranged for this scholarship and who secretly paid the remainder of his bills? Today the cost of attending Yale might be $50,00.00 per year. It was there that he met and married Ms. Hillary Diane Rodham, of Chicago, a Wellesley Valedictorian. After graduation, the newlyweds moved back to Clinton's home state of Arkansas. At that time their friends marveled that these two well-trained lawyers from Yale, who could have immediately commanded high fees in New York or Boston, would elect to return and live in the second poorest state in the country? What these friends did not understand was that Bill Clinton was being "channeled" toward a

---

17   The Holy Bible, KJV Gene. The Holy Bible, KJV Genesi

higher goal than simply making money. Certainly, by this time, and probably since his high-school days, Clinton knew of his true parentage and was aware, also, that if he followed in the path set down for him his future was assured. If he "played the game" he could "do no wrong."

Although he was a licensed attorney, Clinton decided not to enter private practice but rather he took up an instructorship at the University of Arkansas. Then, after approximately two years, he commenced his true career; to run for political office—almost any office. First, he ran for the U.S. Congress and lost. Then, in 1976, with the help of a friend, Jim Guy Tucker, who later became an Arkansas Governor also, he ran for State Attorney General; and won. Thus, he became one of those remarkable attorneys general, as was Robert Kennedy, who have taken on that high office without ever having argued a case in court!

This victory provided a springboard for the "boy wonder." He seemed to have everything necessary to succeed in politics: good looks, organizational ability and the full support of the political "movers and shakers" within the state's Democratic Party. Next, was the Governor's office to which he was elected and reelected, repeatedly, from 1978 through 1991, until an opportunity for the top job beckoned—the Presidency of the United States.

However, before this could happen another hurdle needed to be cleared; approval by the super-secret International Bilderberg Society. This organization was holding its biennial meeting, in 1991, in Baden-Baden, Germany. Membership is limited to the world's highest leaders in government, business and finance. Only members and invited guests may attend. Here it is decided which policies are to be followed and which politicians shall be appointed to rule over their respective provinces. In 1991, one of those rarely invited guests was Mr. Bill Clinton, the Governor of Arkansas, U S of A![18] What transpired during those meetings will never be known; but, obviously, the vetting was successful. Bill Clinton's candidacy was approved and he was told the path was open. His campaign for the Presidency could begin.

During the years (1979–1991) while he personally behaved in a reckless manner by maintaining liaisons with many women, Bill Clinton, then as now, seemed to live in a protective shield within which he could do exactly as he pleased without facing any serious consequences.

This shield of immunity is powerful and real. Those who have attempted to penetrate it or to expose Mr. Clinton's private life or his secret political connections have been thwarted or frightened into silence. Some, who have come too close and who could not be intimidated, have died from "accidents" or commit-

---

18   John Milton. Sonnet XII. The Student's Milton. Appel

ted "suicide" under mysterious circumstances. The plain fact is, whatever his scandals, Bill Clinton personally is untouchable. He lives in a cocoon of steel.

In April 1998 on an official tour through the Union of South Africa, Mr. Clinton was invited to attend a Catholic Mass, and with the other parishioners, he took Communion at the alter rail. This caused a religious flurry within both Protestants and Catholics circles back in the United States. The Protestants asked, how could Clinton, who represents he is a Baptist and who attended the Episcopal Church in Washington, D.C., take communion in a Catholic Church? The Catholics asked, how could he have been *permitted* to take communion, because it is forbidden by the Church to provide communion to non-Catholics? The significance of this act may well be more than it appears to be, on the surface.

Consider that, if he had a Catholic father and attended a Catholic parochial grade school for eight years[19] and then spent four years at Georgetown University, the leading Jesuitical institution in America, could Bill Clinton be anything else *but* a crypto-Catholic? In truth, his taking communion during a Catholic Mass was his right; and the Catholics have no cause for alarm.

If it is true that Bill Clinton is the son of Joseph P. Kennedy; and the evidence points strongly in that direction; positive proof only awaits a comparison of the family's DNA. Of course such a test would be stoutly resisted. But one other point of proof is the dominant personality flaw of this president who exhibits the same inherited pathology of reckless satyriasis; shamelessly exhibited, in the behavior of all the Kennedy males from Joseph P. Kennedy, himself, down through his sons and grandsons. In what other family, rich or poor, has this inherited trait been so blatantly displayed? If he is not a Kennedy, he certainly deserves adoption into the clan.

In comparing the American Empire to that of Ancient Rome one might well conclude that if, in Franklin D. Roosevelt we had our Augustus, and in Lyndon Johnson our Tiberius, perhaps, in Bill Clinton, we have had our own Caligula.

TIME SCALE: The United States (AD 1999) = Rome (AD 37)

MYSTERY WOMAN—HILLARY CLINTON

During the waning years of the Clinton presidency, PAGAN decided that it had, in Ms. Hillary Clinton, a politician capable of being groomed to be America's first woman president and who, in fact, is probably more dedicated to the Grand Design than her husband has ever been.

All during their relationship, she and her husband have had what they call an "open marriage;" a marriage of convenience. They were, and are, married in name

---

19   From "Carl Sandbur

only. She tolerated Bill Clinton's behavior because she knew he was programmed to be President and that she could ride along as his advisor.

In 1999, it was decided by PAGAN, that it was her turn to enter into politics after Bill Clinton's term ended. But, what is the necessary first step? An important step toward the presidency, often, is to first be elected to the Senate.

But, how can one successfully be elected? The answer is to run in a safe Democratic district such as New York State.

But, how do you accomplish that? The incumbent, Senator Patrick Moynihan, was up for reelection in the year 2000 and was so entrenched and so well respected he was virtually assured of re-election. But to PAGAN this presented no problem. They simply told Senator Moynihan he needed to withdraw from the race and retire from the Senate. Moynihan, as a loyal son of the Church, dutifully complied. Thereupon, Ms. Clinton filed for the primary and handily won the Democratic nomination; although with some serious questions about whether the Orthodox Jewish vote in certain religious communities had been bought by the government granting them immunity from their alleged crimes of welfare fraud.

But, during the General Election campaign, Ms. Clinton found herself up against a formidable Republican opponent in the person of New York City's Mayor, Rudolph Giuliani. As the General Election campaign approached to within two months of election day, it became apparent that the popular mayor would defeat Ms. Clinton. Something had to be done. Therefore, as in the case of Senator Moynihan, Giuliani was signaled that he needed to withdraw so Ms. Clinton could win. This, at first, he refused to do.

Therefore, to more clearly communicate its will, PAGAN brought up its heavy artillery and began to bombard Giuliani unmercifully. Newspaper articles appeared describing his extramarital affairs and his current difficulties with his wife whom he was divorcing. To make its will even more clear, a long article appeared in the New Yorker magazine revealing that Giuliani's father had once been convicted of armed robbery and had served a sentence in Sing Sing prison. A young Rudy Giuliani had neglected to report this fact when he applied for employment with the FBI. No doubt he was protecting his family's reputation but this was a serious omission.

Finally, Giuliani realized he had no choice. PAGAN was too powerful and might go to any lengths; including the possibility of assassination. Therefore, Rudy threw in his hand in mid-August of 2000. Citing he had just discovered he had prostate cancer, he withdrew from the race. This surprising announcement forced the New York Republican Party to nominate a "no name" replacement candidate; leaving the field wide open for PAGAN'S chosen one, Ms. Hillary Rodham Clinton, now known as U. S. Senator Clinton.

After the election, Giuliani's health magically improved and, as of this writing, he is making a huge income in speaking fees while traveling about the country. This is how the will of the people is served under the rule of PAGAN.

THE GRAND CONSPIRACY

Jesuitical control of the Clinton administration was fully apparent. Clinton's closest advisor personal attorney (and controller?) was Robert Bennett. This is the same Robert Bennett whose brother is William J. (Bill) Bennett. Bill Bennett is the one-time Secretary of Education and later Drug Czar under the Republican President, Ronald Reagan. Bill Bennett is the one who recently lectured the nation upon the subject of morality by publishing a literary anthology called "The Book of Virtues." Are these two brothers not the right and left hands of the Jesuits; the right and left hands of PAGAN in control of our government? This seems entirely likely.

While the general public does not concern itself with the religious affiliation of its political candidates, the Jesuits and the Catholic Church think of nothing else. They believe wholeheartedly in the fusion of *their* Church with the State. Ordinary Protestant Christians are lulled into an attitude of toleration but in the cities and states with a sizeable Catholic population Catholics vote, en-bloc, and only Catholic candidates can be elected to the major offices. Once elected, they receive the best of press coverage and are lauded as progressive political peace-makers. Examples, at this time, of these rising leaders are: ex New York Governor Cuomo, New York City's ex Mayor Rudolf Gulianni, Michigan's Governor John Engler, Chicago's Mayor Richard Daly Jr. and Governor Arnold Schwartzenegger, of California.

On the national scene, perennial presidential candidates such as John McCain, Alan Keyes, Patrick Buchanan, and John Kerry are also Catholic. All, all, are honorable men but, if elected, they ultimately would be under the control of the Jesuits and the Vatican.

Without Americans noticing, this Catholic cabal has gained nearly total control of the Executive and Judicial Branches—while to cover its tracks and to maintain the support of the Zionists, some high Cabinet and diplomatic posts are reserved for their members. These often include the office of Secretary of State, Secretary of Defense, Secretary of the Treasury, Chairmanship of the Federal Reserve Board and others including the Directorship of the Internal Revenue Service.

During a recently televised interview, the author of a new book entitled "My Spy," gave an account of her lifetime as the wife of a high-level career CIA officer. As an aside, during the question and answer period she mentioned that in the countries where she and her husband were stationed, the CIA commonly works with the local the Catholic priests, employing them as informants and as recruiters of new agents. For convenience, the CIA officers and the priests often conducted their meetings in the privacy of the confessional booths! Naively, this woman thought it was permissible for her to make these facts known because, as she explained, while her husband had been sworn to secrecy; she had not been. If these statements are true, and it would seem that the woman was giving first-hand testimony, this revelation is appalling. It means that Catholic priests, world-wide, are cooperating with the CIA in what is actually espionage within those sovereign countries. This must mean, also, that these priests' acts of espionage are being carried out upon the direct orders of the Papacy. It means, further, that the rivers of tax money now given annually to fund the work of the CIA are really being spent on behalf of a foreign power, the Vatican. (Kiyonaga, Bina Cady "My Spy" Avon Books 2000)

Our Department of Defense (Offense) is at the Vatican's beck and call. It effectively provides PAGAN with an Army, Navy and Air Force. The disgraceful spectacle of the U. S. Army, Navy and Air Forces, waging undeclared wars against smaller nations in so-called "peacekeeping" and "anti-terrorist" missions results solely from the Vatican's need to attain its political goals by chastising all peoples and all governments not yet submitted to its rule. (i.e. Serbia Afghanistan Iraq etc.) Truly, the American people should not be concerned with whether our political candidates or government officials are Republicans or Democrats; or whether they are of the "right" the "left or of the "center;" the question should be:—what is their religion?"

In this connection one may ask where, physically, is this Catholic Cabal located in America? The answer is that it lies in plain sight. It is within Georgetown University, in the District of Columbia, where the Jesuits continuously supply new cadres of brain washed "public-servants' who graduate to staff the highest offices in our government agencies.

CHAPTER NINE

PAGAN UNMASKED

While there may appear to be a whole spectrum of anti-Christian groups that may collectively be called PAGAN, at their head is the Second Roman Imperial Government lodged in the Vatican in Rome. Movements commonly viewed as non-Christian such as Communism, Freemasonry, Mormonism, Zionism, B'nai B'rith, Jehovah's Witnesses and the New Age Cults are, in reality, front organizations; movements invented and supported by the Papacy over the past five hundred years. All are controlled, ultimately, by those Masters of Deceit; the Society of Jesus. This Second Roman Imperial Government *is* the Great Conspiracy!

It must be understood by Biblical Christians that the Vatican **is a** conspiratorial imperial political organization and not a Christian church! Its religious operations are only a sham; carried on to retain the loyalty of the common believers who are its foot soldiers and who are led about by the use of mock-ceremonies, superstitious worship of idols, holy relics, dead saints, shrines and false dogmas. Since its founding, it has always been first and foremost an international political organization It is the Old Roman Empire alive and well today. It now seeks to conquer all nations and all peoples and to bring them under its rule. It masks itself as a church only to conceal its true mission. What other organization has a leader who claims to be "infallible." Who else but the Pope, they will assert, has the God-given right to rule New World Order?

Michael Sindona, an Italian lawyer and financier, who had unpleasant dealings with the Vatican Bank, had this to say:

"I know now that the power of the Vatican is the system of time. We die but it does not. A lifetime is nothing to the centuries that are the slow beat of the Vatican's pulse. They condemned Galileo and they are still trying the case three hundred years later. It is the system of time. Such men as Cardinal Marcinkus are but cogs; replaced as they wear out every half-century or so. It is terrible." ("Power on Earth," Nick Tosches, 1968, Arbor House, New York: p.218.)

MYSTERY MEN

Beware of the "mystery men"—those who slip into your midst and who then take charge of your group or movement. In a former time as a member of a "patriotic" organization this author had occasion to know one of these. The local organization was made up of sincere "believers;" but as it grew larger it required a full time paid administrator. However the members were not permitted to choose one of their own for the position but rather the national headquarters insisted on appointing a stranger. He soon arrived and took up the tasks of leadership

This man seemed to be highly qualified. He was about thirty years of age, handsome, and an excellent public speaker. What more could one ask. However we thought it curious we could learn nothing about his background or his personal life; his friends or family. He made the required appearances at scheduled meetings but thereafter disappeared for considerable periods of time. When asked where he lived he was evasive but at one time said he lived with his mother. This was found to be untrue.

It was strange also that while he was young and unmarried he seemingly had no connection with anyone of the opposite sex. He presented no open signs of homosexuality; so what could explain the character of this paragon of accomplishment and virtue?

Suspecting that "what is too good to be true generally is *not* true" one of the members brought his suspicions to him and he confessed, finally, that he was in reality a Jesuit priest assigned to act as our controller. His *home* was the local Jesuit seminary. The members now knew the truth—but only too late. For over four years this secret Jesuit agent had led our organization into a series of useless activities that totally neutralized our efforts. With his identity exposed, the organization collapsed and its members dispersed.

Consider the implications of this revelation. It indicates that the Jesuits' network is so extensive it is able to monitor and to control political activities down to the local level. It means PAGAN may, at will, send one of its agents to infiltrate and take over any potential opposition group.

THE REFORMATION

While the Protestant Christian Reformation did not begin until 1517 it was preceded by the efforts of the Albegensians, Paulicans, Lollards, Hussites and Primitive Christians who for a thousand years had nurtured their faith through terrible trials. These movements were amalgamated by the efforts of John Wycliff, Martin Luther, John Calvin, John Knox and by the publishing of the Bible; Luther's in German and William Tyndale's in English. Then, because many of the rulers in Northern Europe found it desirable to throw off the political grip of the papacy, England, Prussia, Holland and the Scandinavian countries soon embraced Protestantism.

Infuriated, the Vatican responded by organizing a vast War of Counter-Reformation against those Protestant states which they deemed to be heretic. During the latter half of the 16th Century they directed the Hapsburg monarchy the rulers over Spain and Austria, the richest and most powerful Catholic areas of Europe, to wage a war of extermination against the Protestants of the Netherlands. This is why during this same period in 1588 the Spanish Hapsburg (Phillip II of Spain) sent his Armada against England. Its purpose was not to defeat the British Navy but to avoid it. The plan was to land a Spanish army on England's South Coast; to, then, conquer England and reestablish Catholic rule. They planned to seize Queen Elizabeth and cut off her head. In the view of Philip II she was no Queen at all but a Protestant usurper. In the event, however, the superior seamanship of the English fleet under Sir Francis Drake prevented the Armada from landing and kept the Spanish invasion from being successful. Had Phillip's army landed ashore there was little in England that could have opposed it.

In the 17th Century, beginning in 1618, the House of Hapsburg, the Catholic rulers of both Spain and Austria, began what became the "Thirty Years War" against the smaller Protestant states of Bohemia, Central and Northern Germany. This series of campaigns devastated Central Europe but resulted basically in a military stalemate. At the Treaty of Westphalia, signed in 1648, the lines on the map were finally drawn between the Protestant and Catholic realms in Europe that define their borders today.

The Civil War in England (1641-1647) was waged at the same time as the Thirty Years' War on the Continent. The Protestant victories of Naseby, in 1645, and later the Battle of the Boyne, in 1690, meant that the Jesuits once again had failed to recover Great Britain by military means. Frustrated, they resolved to continue their struggle but by means of subversion. It was then that they invented Accepted Masonry as their primary tool with which to deceive their enemies.

ADAM WEISHAUPT

By the year 1770 the Jesuit's meddling in the political affairs in France, Spain and Holland caused those countries to petition the Pope to curb the Jesuit Order. In response, In 1773, Pope Clement XIV revoked the authority of the Jesuits and took over their properties in Rome. But, while the Jesuits appeared to obey, they, in fact, went underground; continuing with a new form of subversive organization. It was then that they began to operate as the Illuminati. The plan was implemented by one Adam Weishaupt with his publication of a document he called the "Constitution of the Order of the Illuminati." Beginning in 1775, Weishaupt organized and took control of a network of Masonic super-lodges which he described now as Illuminated. With these, he intended to train up a cadre of wealthy and militant converts who would eventually conquer the world by infiltrating all of the world's governments. His method was to surround the nominal rulers of each nation, whether they be Kings, Princes or elected officeholders, with his secret agents. He and his successors thus would become the actual rulers of the world while operating from behind a screen of secrecy. As fantastic as this plan then seemed, it has now come close to fulfillment. This is what we call today "The Grand Design."

Modern historians pay slight attention to the life and character of Adam Weishaupt; pretending that the Illuminati movement was long ago and of short duration; that it ended when outlawed by the government of Bavaria in 1785. They imply that Weishaupt was a lone madman who was, finally, reined in by the civil authorities.

But who was Adam Weishaupt—really? The answer is that before he began his organization he was a mere 25 year's old Jesuit priest; a junior instructor in Canon Law at the University of Ingolstadt, in Bavaria, when, quite suddenly, in 1773, he was promoted to a full professorship at that unusually early age. Two years later on May 1st 1775, (the birthday of Satan/Baal?) and seemingly without any preliminary development, he issued his document entitled "The Constitutions;" an organizational plan for the regulation of the Order of the Illuminati. He began to recruit followers from the ranks of the German lesser nobility and from members of ordinary Masonic Lodges. In 1778 he boasted that by then he had enlisted over 1000 followers. Interestingly, all during the 10-year life of the Illuminati movement, 1775-1785, Weishaupt remained a full professor in good standing on the faculty of Ingolstadt University although the nature of his activities must have been well known. The only explanation for this tolerant attitude by the University must be that he was still a Jesuit priest under orders. He was not dismissed from the University until the Bavarian political authorities exposed the Illuminati in 1785. Even then, he was protected. After the official suppression of the Illuminati he suffered no punishment and indeed was offered a pension by the University and given personal asylum at the Court of the Duke of Saxe-Gotha; a brother Illumine. After this he disappeared from the pages of history but presumably continued to direct his followers from, underground, until his death in 1830. (Encyl. Brit. 11th Ed. Vol XIV p.30) It is reported that he "returned" to the Church. (but of course he had never left it.). Presumably, he lived out the rest of his life under an alias; sheltered within some Catholic institution. During the interregnum, the Order continued to operate in Prussia and in the court of Russia. Following the end of the Napoleonic regime, the Order of the Jesuits was restored to favor once more.

From the import of this story what becomes apparent is that the Order of the Illuminati was, in fact, a continuation of the long-standing Jesuitical plan to rule the world by political subversion. It indicates that Adam Weishaupt SJ, (Ibid. p.30.) was all the while a leader in advancing this scheme.

A French Illumine' Honore de Mirabeau translated and copied the "Constitutions" and imported them into France under the title "Letters from Arcelsius." Those plans, when put into action by the Lodges of the Masonic Grand Orient of France, undermined the French monarchy and fundamentally set off the Revolution.

Outlined in the "Constitutions" was a fall-back position, whereby, if the Brethren were not able to operate openly they could always continue their work underground by posing as professors, "Literary Societies" or by organizing student Fraternities that could call themselves such names as "The German Union" or as the Order of The Scull and Bones; with Lodge # 1 in Heidelberg Germany and with Lodge # 2 at Yale University, USA. to convert young scholars into accepting their "advanced" ideas. [20] Many famous men of the time were claimed as members. The list included Goethe, and Herder. In the 19th Century these "new ideas" surfaced as the Romantic Movement. One example is the tale of "Dr. Frankenstein," authored by Mary Shelly. "Frankenstein" tells the story of the attempt by a fully "liberated" scientist to become a "creator" of life in defiance of God; a precursor to the Theory of Evolution. To be sure, the attempt was flawed—but it was partially successful. The creature, after all, did live for a time; did it not?

The Order Of The Illuminati is certainly not dead. It lives on as PAGAN. The Jesuitical plot that we now call PAGAN is the evil root of that plant whose branches we mistakenly view as unrelated political and social movements such as "Freemasonry," "Mormonism," "Liberalism," "Zionism," "Progressivism," "Socialism," and "Communism." The common thread that binds them all together is their denial of God and their rejection of the Gospel of His Son, Jesus Christ.

The Spanish Inquisition lives on in the form of the Society of Jesus It is not a relic of history; it is a virulent and present threat; it is an active conspiracy advancing toward its goal of world dominion without regard for honesty, morality or truth. This Jesuit Gestapo is the most dangerous organization on Earth. Their Inquisition has surfaced once again and can be recognized, today, in the tortures being inflicted upon the helpless prisoners of Abu Garib Prison, in Iraq, and in our own government's gulag in Guantanamo, Cuba.

---

20 Ibid.

THE FRENCH REVOLUTION

The Jesuit's power over the government of 18th Century France, by working through Masonry, was open and undoubted. The King of France was but a figurehead. The FreeMasons had taken over control of the government. This is why, in 1778, France decided to finance the American Revolution. It was because the American leaders were brother Masons. Without that vital aid, the American Revolution would not have been successful. In the same vein, most do not see the influence of the Jesuits in the creation and manipulation of the French Revolution. Many of the revolutionary leaders were Jesuits Illuminees or products of a Jesuitical education. Voltaire graduated from a Jesuit college and thereafter spent a lifetime speaking and writing in favor of "reason" and against civilized manners and true religion; which he ridiculed as superstition. Near the end of his days he became an open Freemason; joining the Lodge of the Nine Sisters in Paris. It is reported that he was led down the aisle on the arm of his friend and counterpart, Benjamin Franklin, the American Ambassador to France.

Charles Maurice de Talleyrand was originally the Bishop of Autun. When the revolution began he left his bishopric and became one of the most prominent political controllers of his time.

Elected as a church delegate to the First Estates General, he used his influence to modify acts directed against the Church. His Jesuitical identity protected him during the Terror when nearly every other leader was beheaded. After the Terror, he served as Foreign Minister of France under three regimes; the Consulate, the Napoleonic Empire and the restored Bourbon Monarchy.

Mysteriously, whatever the political climate, Talleyrand always floated to the surface. After the fall of Napoleon, he was appointed the French delegate to the Congress of Vienna by the succeeding Bourbon Monarchy in 1814-15. Afterwards, he continued to perform important roles in political affairs until his death in 1838.

While the French Revolution has been considered by many to have opposed religion this was not quite true-; it was only anti-Christian. With regard to Catholicism, the revolutionaries were quite tolerant. The priesthood was merely required to acknowledge the Government as the supreme political authority and to take an oath of loyalty to the state. While the factions operating within the Revolution were diverse, as events continued on, the hand of the Papacy began to intervene.

The Revolution was first led by radical idealists but it ended under the control of the Church. In 1802 to legitimize his crowning of himself as Emperor, Napoleon signed a "Concordat" with the Pope that, in political terms, meant he had submitted himself and France, as well, back under Papal authority.

During these confused times, Freemasonry received much of the credit (and much of the blame) for the excesses of the Revolution; although, unknowingly, it was only a front group. Nevertheless, its apparent success strengthened Masonry to the point that behind the scenes it has become a major underground force in world politics. Ostensibly opposed to Catholicism it is, in actuality, the Papacy's controlled opposition.

During the first thirty years after the Congress of Vienna Freemasonry, (Illuminism), and its allied movements in Europe fathered Socialism, hired Karl Marx to write the Communist Manifesto and fomented the European Revolutions of 1848. Later it was involved in the Paris Commune of 1870 and in the Bolshevik Revolutions in 1905 and 1917. Communism, essentially, is only a thorough-going form of Masonry; a fact always recognized by the anti-Communist factions in Russia.

Freemasonry and Communism are covert agencies of the Papacy. This fact is confirmed by recognizing that during the seventy years of autocratic Soviet rule in Russia, while religion was outlawed, it was not Catholicism that was outlawed; it was only that ancient enemy of the Catholic Church in the Slavic world, the Eastern Orthodox Church that was outlawed. In Catholic-dominated countries such as Poland; when in 1945 the Communists took over the government, the Church was tolerated and its properties left untouched.

NATIONAL SOCIALISM

If Communism is the left hand of the Papacy National Socialism is its right.

Few remember that the "National Socialistes Deutches Arbeitern Partei" (NSDAP), or NAZI movement, originated and grew to power in the heavily Catholic region of Bavaria—specifically in the city of Munich.

Ambitious for a political career, Adolf Hitler, who was born and educated as a Catholic, joined and then left several political parties before starting his own. The first party he joined was the Catholic Center Party of Austria.

Hitler and his group were not refined intellectuals but rather discharged front-line World War I soldiers who believed they had been betrayed by political traitors at home. They believed that, while they courageously held their positions at the front, Communists and Jewish revolutionaries had betrayed the nation.

It must be remembered that the leaders of these Socialist and Communist revolts such as Rosa Luxembourg, Karl Liebknecht and Carl Eisner *were* all Jews. Considered traitors, they were used by the Nazis to later justify their anti-Semitic policies.

The Nazis formed a party of *direct action*. In "Mein Kampf," Hitler boasts that during public mass meetings, when unable to convert their enemies by oratory, they would club them into agreement. Hitler's whole argument was that "The time for intellectual debate had passed by and that force was the only answer to political questions." With hindsight one sees how this reliance on "force" carried on to extreme, as the ultimate answer in political argument, led to the catastrophe that followed.

As a party of bullies, Nazism was ripe for control by more sophisticated hands. Having had a Catholic education, Hitler admired the organizational principles of the Church—especially the requirement of celibacy for the priesthood. (this is why, as a self-appointed priest, he never married until his last day on Earth) He felt celibacy ensured that an organization's leaders must remain loyal to it because they would have no other legitimate outside relationships—and further, that its leadership must always be replaced from outside—eliminating nepotism; constantly renewing its strength. (Hitler Adolf "Mein Kampf." The Riverside Press 1933 p. 178.)

Joseph Paul Goebbles, the Propaganda Minister, as well as Heinrich Himmler, leader of Hitler's bodyguard who organized the Homeland State Police, called, in German, the "SS" or "Gestapo," were both Catholics. Himmler openly used Jesuitical methods of administration. The following is from a history of the SS:

> "The SS organization had been built up by Himmler on the principles of the Order of the Jesuits. The service statutes and spiritual exercises prescribed by Ignatius Loyola formed a pattern that Himmler assiduously tried to copy. Absolute obedience was the supreme rule. Every order had to be accepted without question."[21] Himmler even instituted an SS Order of Knighthood—with appropriate pagan rituals which were celebrated annually at the Gestapo's "Monastery" Schloss Wewelsburg, near Paderborn, in Westphalia. (Manvell Roger "SS and Gestapo" Ballentine Books 1969 p. 40.)

The degree of Jesuitical influence during subsequent events is difficult to trace but it is obvious that the Nazis were in agreement with the Jesuits upon one essential principle; "The end justifies the means."

It is impossible for an outsider to comprehend how the German people could have elected Adolph Hitler and the Nazi Party into power, in 1932, until one considers the situation the German people felt they faced. To one side were the ineffective leaders of the Weimar Republic who had failed to solve the nation's economic depression and who lacked the will to maintain civil order; and on the other side, a para-military Communist Party, financed by Soviet Russia, that was making repeated attempts at an armed revolution. By voting for Hitler, the peo-

---

21. Kessler, Ronald. "Sin

ple believed they were voting for a man of the center; voting for a party and a leader who could solve both these problems.

Initially the social problems were solved. The Nazis reformed the economy and the Communist threat was eliminated. But eliminated also were the people's civil rights and any personal protection under a rule of law. By 1939 with the dictatorship complete, Germany had achieved "order" but at a price; the price of political slavery.

Looking back upon that summer of 1939 mysterious forces were at work urging Hitler to take Germany into war. Had he stopped his aggression at that point, Germany and Europe might have been spared. The national goals had been achieved with the adjacent German speaking peoples of Austria and Czechoslovakia repatriated into the Reich. The economic depression had been overcome and, with a conversion to a "peace policy" and with a relaxation of the civil dictatorship, Germany could have assumed its natural role as the most economically powerful nation in Europe. A war for territory was unnecessary. Poland and Russia, were, and are, economic sinkholes; not worth the life of one single German soldier.

As to the proximate question of the Danzig Corridor, the answer was that Germany could have bought the corridor from Poland or the right to build an elevated autobahn over it for a single boxcar full of gold.

The final temptation of Hitler was the offer from the Soviet Russians to make an agreement to jointly invade Poland and divide it up equally. This offer was accepted by the signing of the German-Russian Non-Aggression Pact of August 1939.

Who could have brought off this unbelievable diplomatic reversal except the Left and the Right hands of PAGAN? On the German side that force must have come from the Society of Jesus and from the Russian's side, International Freemasonry.

As Hitler was deciding for peace or war, this signal from the Russians must have appeared to be an unbelievable green light. It would temporarily protect his rear on the Eastern Front. It tipped the scales toward war. To that moment, the presence of a hostile Russian Army in the East had always checked any serious German thoughts of aggression against Poland. Had this remained the case, the war would not have begun.

Six years later, at war's end, Germany lay prostrate with many of its cities in ruins. But the damage was not equally shared. Southern Germany, Bavaria and its cities, were nearly untouched; while Northern Germany, i.e. Protestant Germany, was devastated and overrun. Northern Germany was afterwards occupied for forty-five years by the powers of Soviet atheism. Do we not see the hand of the Papacy here? Was not World War II just one more episode in the Vatican's six hundred years' long ongoing war of Counter-Reformation against the followers of Jesus Christ? Is the Grand Design the War of The Counter-Reformation in disguise?

Throughout the terrors of Nazism, as whole classes of persons deemed threats to the state such as Jews, Gypsies, homosexuals, or militant Christians were systematically imprisoned and killed, there is no mention of even one Catholic priest or bishop having suffered a similar fate. Catholics were a protected class.

In a recently published book "Hitler's Pope" The Secret History of Pius XII" the author John Cornwell (Cornwell, John "Hitler's Pope" Viking Penguin Publishing Co. 1999 New York p. 110.), describes the Vatican's alliance with the National Socialist Movement.

Cardinal Eugenio Pacelli (later to be named Pope Pius XII), had been the papal Nuncio to Germany (i.e. papal ambassador) from 1919 to 1929. Pacelli was a highly trained canon lawyer who had previously been the Vatican's Under-Secretary of State. In that position he had edited the present Code of Canon Law, issued in 1917. This Code is the legal capstone of the papal reorganization process begun in 1870 with first modern Vatican Council, Vatican I (the Council that promulgated the Dogma of Papal Infallibility.)

For centuries, (and the Vatican does think in terms of centuries), The Catholic churches of Germany, France, England, Ireland and the Americas had enjoyed a degree of independence. While looking toward Rome for doctrinal guidance their distance from Rome meant they were largely self-ruling with regard to the appointment of priests and bishops and in the operation of their parochial schools. However, with the rise of strong nation-states, governments, like Germany, often assumed a veto-power over the naming of new priests and bishops within their countries. In Germany the establishment of the government's authority was described as a "Kulture Kampf" (a culture battle) waged in the late 19th Century by the government of Chancellor Bismark. This meant, in essence, that there were several Catholic Churches outside of Italy; the German, the French, the Irish and the American etc. It was this diversity that the Jesuits set out to destroy.

Seeing that the means of communication were improving, the Vatican's new policy became to somehow negotiate agreements with each of the national governments to permit the Vatican to exert direct and sovereign control over their own Catholic organizations, priests, bishops, and schools in every country; without interference from the civil powers. The overriding goal of the Vatican Curia was and still is to weld these various churches into one single worldwide Church-Empire, directly controlled from Rome, and forged them into one pyramidal structure; with the infallible Pope at its apex. While stationed as a papal Nuncio in Germany from 1919 until 1929, Cardinal Pacelli had as his settled goal the implementation of this policy for the German Catholic Church.

As the Nazi Party grew throughout the 1920s it was opposed by the lay Catholics of Germany. Catholic newspapers warned their readers not to be deceived by the Nazi's godless programs. The Catholic Center Party (the

Zentrum) was the largest party in opposition to Hitler and it held enough voting power in the Reichstag to have prevented legalization of the dictatorship. Had the Zentrum been permitted to stand firm, Adolf Hitler would never have become a legalized dictator.

But, upon the naming of Hitler as Chancellor in 1932, Pacelli sensed a weakness in Hitler's position. He believed it was the moment to approach the German Government to get what he and the Vatican desperately wanted; direct control over the German Catholic Church. To achieve this, Cardinal Pacelli, now as Secretary of State of the Vatican, personally negotiated the "Concordat" (a treaty) between the Vatican and Germany that was signed in 1933. By this treaty, it was agreed that, as a quid pro quo for absolute control the Catholic Churches in Germany, the Vatican would, on its part, order its priests its newspapers and the Catholic Center Party to cease all opposition to National Socialism. In effect, Pacelli and the Vatican placed Hitler on his throne. That is why without Catholic opposition the "Enabling Act of 1933" could be passed by the Reichstag. This Act granted Hitler dictatorial power—legally.

By loosing Hitler upon the world, Pacelli and the Vatican were the primary cause of the Second World War. But the death of millions seemed not to matter to the Vatican. If they could achieve just one of their policy goals, the mass destruction of men and nations was, to them, of little concern.

The author of this work, John Cornwell, is himself an English Catholic and he states that his purpose in researching for this book was to defend the memory of Pius XII from the charges that had been made against him that he and the Vatican did little to curb the Jewish Holocaust during World War II. But to the author's "moral shock," he found that the suspicions were not only true but that the evidence was even more damning. This forced him, in all honesty, to write a far different book from the one he had originally planned.

The Vatican recently held a conference to consider making an "Apology" for the mass-murders carried out by the Inquisition from the 13th to 19th Centuries. But next, they admitted that the Inquisition may still be in operation. Pope John Paul II. Stated that: "In the case of the Nazi Holocaust some of its "children" had erred; but not the Church itself!" (Associated Press November 1 1998) Who were those "children;" who had erred? Were they not Adolf Hitler, Heinrich Himmler, Joseph Paul Goebbles, Von Ribbentrop and the rest?

What we see now, is that in Germany during the period 1930-1945 what occurred was that the Vatican and the Catholic Church of Bavaria formed the Nazi (Catholic) Party and seized power over the whole of Germany. They, then, used that nation's industrial and military might to conquer and to mobilize the rest of Europe in preparation for the great drive to the East. (Die Drang Nach Osten) The German invasion of Russia was designed to carry out the will of the

Papacy whose eternal goal has been the destruction of its mortal enemy, Russia along with its Eastern Orthodox Church. This is why millions of German soldiers both Protestant and Catholic, and millions of Russian soldiers as well, were forced to slaughter each other on the frozen steppes of Central Asia. That is why Northern (Protestant) Germany was bombed into rubble. There can be no other answer. The triumphant Germans of 1940 were the masters of Europe and had no sensible reason to extend their conquests any further into the primitive vastness of Russia.

Now also, we can understand as a parallel event; why the Emperor Napoleon made his invasion of Russia in 1812. It was an unnecessary campaign made only to carry out the schemes of the Pope. Upon becoming Emperor of France in 1802 Napoleon had, like Hitler, signed a Concordat with the Vatican and from that time on he was its lackey.

The Jesuits and the Illuminati controlled the Grand Orient of France and Napoleon was a committed Mason. Their will was his will no matter how many French and other nations' soldiers had to die in the snow during his disgraceful retreat from Moscow.

Is another assault upon Russia now in preparation? With the Vatican and its dupes in effective control of the United States government, can we not recognize that the United States Army, Navy and Air Force have been mobilized and moved up to the Southern borders of Russia in Afghanistan, Iraq, Kazakhstan, and Turkey? The people of the United States do not understand why they have been sent there; and they do not want them there. Nevertheless, our troops have been sent into these countries upon orders issued from the Executive Branch; orders that are clearly unconstitutional. One spark, one single arranged incident, could be used to justify the beginning of World War III because the President is now able to declare wars at his pleasure; or rather at the pleasure of the Pope. What might it cost in American lives and treasure, to wage such a vast campaign? It could impoverish or destroy America. But who cares? If it is the settled will of the Pope, "Dieu le Veult." (God Wills It!)

INTERNATIONAL SOCIALISM

In countries with a large Catholic presence such as Italy or Southern Germany, Jesuitism influences political developments, openly. In non-Catholic countries it operates through front groups to conceal its identity while still carrying through its programs. Such was the case in the Soviet Experiment wherein the ideals and the terrorist tactics of the Inquisition were applied against the Russian peoples with a vengeance.

The reigns of both Lenin and Stalin were an unending series of pogroms conducted to exterminate any class or individual whom it suspected might be a threat to the state. Their aim was to obliterate all memory of the former Czarist social system and its accompanying Eastern Orthodox Church.

Standing where we are today is it not possible that both the Nazi and the Communist uses of terror were in reality modern continuations of the Spanish Inquisition visited upon the populations of Christian nations across the length and breadth of Europe from the English Channel to the Urals?

What has been the result? The result of these terrors has been the crushing of true Christianity in the hearts of many survivors. These have fallen into the skepsis of relativism wherein all faith has been lost. These have become fertile soil for the growth of a new religious solution; Ecumenism.

Do not for a moment believe that in any Ecumenical Reunification, the Roman Catholic Empire intends to give up one single inch of doctrinal ground. The Second Roman Empire, by its own assertion, is infallible and unyielding. The following is from an Encyclical of Pope Pius XI issued January 6, 1928. It reads:

> "First the perfection of the Catholic religion as a system of doctrine and practice has been defined with increasing clearness and rigidity through the centuries. The system itself is super-naturally perfect, and is now exactly what it was when it was first revealed by Christ and what it must always be. The Roman Catholic Church therefore has nothing to learn from any other source. It cannot compromise or co-operate where such action would cast any doubt upon its assurance of its own perfection."
>
> "Second, is the infallibility of the visible church as the agency to which God has entrusted the preservation, definition, administration and propagation of that perfect system; (This is represented) specifically in the infallibility of the pope as the head of the church and the personal representative of Christ." ("The Vision or the Degrees of Glory Ibid. p.110)

This "Dogma of Papal Infallibility" is a late addition to the Catholic schema, not adopted until the First Vatican Council was convened in 1870. (Vatican I) Previously, Catholicism was a common set of beliefs shared by the church membership with the Bishop of Rome only as its symbolic head. Catholic bishops worldwide retained their powers of independent thought and opinion. When they were summoned to attend the Council the bishops were first told it was only to discuss improvements in the practice of the Catholic Church; adjusting it to the cultural changes in the world since the last Council, The Council of Trent,

1545-1563. However, shortly before they convened, it was revealed that the Society of Jesus was proposing that the principal work of Council should be the proclamation of "Papal Infallibility" as a new dogma.

Learning of this, the bishops from Germany, France, America and Great Britain voiced strong opposition. Nevertheless their protests were overcome and, as many said at that time, they made the "sacrifice of reason" and voted "with bleeding hearts" to avoid a schism. Under protest, on July 18th 1870, they accepted "Papal Infallibility" as a dogma. The following is an astounding paragraph, describing Vatican I, quoted verbatim from an article on the Vatican to be found in the Encyclopedia Britannica of 1911.

"By implementing this Dogma the Jesuits effectively seized control over every subsequent Pope; converting him from being the spiritual head of the Catholic Church into an absolute dictator of an independent nation, The Vatican State. As ruler of this State, the Pope pays no taxes, answers to no Parliament or to any Supreme Court. From his pronouncements, dissent is impossible. With the adoption of this Dogma, the Jesuits have welded the Papacy, the Vatican and the whole Catholic system into a disciplined army with which to effect the psychological and political conquest of the world." (Quotation) (Encyclopedia Britannica, 11th Edition 1911, Vol. XXVII p. 949 et seq.)

Here revealed, is the "Grand Design:" The New One World Dis-Order based upon a single One World Universal (i.e. Catholic) New Roman Empire. If successful, this Jesuitical plot will bring down upon humanity another "Dark Age" just as it did upon Medieval Europe. The whole world's cultures and economies will be frozen into place; or worse regressed, backward, into Feudalism; into a system, fixed and unchangeable, with new ideas being deemed heretical if they threaten the static rule of the Papacy.

As in the time of Constantine, now comes the greater tyranny; the tyranny over the mind—a new Inquisition—the Jesuit's political dictatorship.

TIME SCALE: Rome 350 AD = World Culture 2050 AD.

CHAPTER TEN

COUNTER-ATTACK

Because we can now see that PAGAN is the collective of the world's anti-Christian religions we must arm ourselves with that understanding and go forth to wage a spiritual war against this evil; to drive PAGAN back into the mouth of Hell from whence it came.

Some may be offended to learn that their church or organization is an agency of PAGAN. We do not condemn the individual members of any religious group but do encourage them to separate from those organizations which, unknown to them, compromise with the poisonous doctrines of the Anti-Christ. The hour is late. The realities of our time must be addressed. We trust in the charity of our readers. God knows our hearts and will make the final judgment.

Be uplifted! It is an absolute certainty that just as Christ's doctrines turned back the powers of PAGAN when it ruled as the Old Roman Political Empire so too will it save the world from the social torments of our time. The answer for us is not to waste our energies attempting to reform a corrupted political system but to instead practice "Christian Separation."

While we must work and live in the everyday world, Christians should separate themselves from evil by stepping over into the "Kingdom of God." Make no mistake this spiritual "Kingdom" does exist; it exists 'round about us at this very moment. If you hear and understand its call, then, by taking one emotional and intellectual step, you will find the open door and your place of perfect safety.

How may this kingdom be approached? Unbelievable as it seems to skeptic minds, God and His Kingdom may be contacted by the power of prayer. This is how it happens. The petitioner (that is you or I) humbly *submits* his mind and will to the supremacy of God and asks for mercy; and for His help with our most pressing personal problems. Shortly, within days, God responds with an entirely unexpected arrangement of events that provides a solution, far more perfect, than was originally imagined to be possible. Try it—you will like it!

I cannot explain "how," but prayer works. There is a saying that: "Prayer does not change God, it changes you." That is possibly true, but there is more than that. The very fact of the response to prayer indicates there is something else, something beyond the mechanistic scientific explanations. According to current thought, should the efficacy of prayer exist; (i.e. your speaking to God and God responding to you) this would mean "action at a distance" a fact whose possibility is denied by the logicians of science. Even though, "action at a distance" can be confirmed by scientific demonstration, its possibility is denied by modern "scientists" for the very reason that should "action at a distance" be proven to be true, it would invalidate their pseudo-scientific religious dogma that "science" can find no God. They dare not believe in action-at-a-distance because they wish find only that which they seek. However, it is a matter of clear fact that gravity and the whole electromagnetic spectrum, from atoms to electricity, to radio, and to light all demonstrate "action at a distance" and nothing else.

Is "action at a distance" unscientific and absurd? Consider the reality of our situation on this planet drifting alone in limitless space. Think of the functions of our physical bodies which operate perfectly for three score years and ten without our giving any conscious thought to the millions of electrical and chemical reactions that regulate our existence. Do we consciously direct these functions? The answer is no. Therefore, another force must be at work. What is it? From the bacteria and the viruses to mammals and to man, an unknown force purposes and directs us all. To give it a name it is "Providence"—it is the Will of God. In our present lives we have no better answer but in the after-life we will be told and we will understand. Be assured there is an after-life for, in some manner, we live on in another form.

This present writer had a glimpse of this state having once had what is now called the "after death experience." At 39 years of age I became ill with a condition that could not be diagnosed by my doctor but which I later discovered was probably a severe lack of potassium. One night, feeling extremely affected, I lay down to rest but in that sleep, I passed over into what I at first interpreted to be a dream but which I soon concluded was not a dream at all but something other—something real!

This was the state of being fully awake but not in my own body. As many since have described it I was floating forward in a black tunnel which appeared to be narrowing in the form of the interior of an old-fashioned camera bellows. It was very long but narrowing. At the farther end I saw a light which, as I drew closer, became the lighted face of a beautiful being framed in a window; a being which I recognized, somehow, from the past and from whom I had long been separated. It was radiant. It was the face of perfection.

Was it the androgen, the perfect man/woman, the divine? Christ? I could not tell but it welcomed me and it radiated love. To the left an archway opened from which streamed a golden light. It drew me on. I knew this was the way but I knew that passing through that door meant no return. Reluctant to go as I was still responsible for a young family, I asked to be returned. This request was instantly granted, and I found myself shaken but alive back in my own bed. Then, because I had never heard of this experience from others, I kept it a personal secret for many years; thinking no one would believe it.

A further aspect of this experience was that as I drifted toward the light and realized I might be dying, I asked the Great Question; "What is the meaning of life and the purpose of the universe." With that, the answer flowed instantly into my mind. It was what I had always known but which I had somehow forgotten when born into human form. It was as if I had had a locked file in my memory that was now reopened. Returned to life, I cannot remember what that answer was but I can say with certainty the answer explained "All" it explained the Glory and Purpose of the Universe which is wonderful, ordered, perfect, sure, just and soul-satisfying; it was the answer to all questioning.

Because the political scene is now subject to the dictates of PAGAN how can its evil be opposed and made impotent? The answer is; because it is a system of mind-control it must be defeated by the one force it fears the most, the opposing belief system represented by the Word of God and the Gospel of His Son, the Lord Jesus Christ.

How does one apply this weapon? Not by organizing into political movements; for being the masters of "organization" and with armies of secret informants at its disposal PAGAN will infiltrate your organization by the time it has grown to half a dozen members. No. Organizing on a large scale is not the way. How then? The correct method is that each of us must work at this task by converting others; one by one This requires that you and every militant Christian return to the fundamental teachings of the faith so brilliantly expounded in the writings of the Apostle Paul. These are found in their original purity only in the King James translation of the Bible.

The ONLY true and authoritative bible, in the English language, is the King James translation. Testifying to its value is the fact that it is becoming a "suppressed" book.

When visiting a major bookstore you may notice that in the religion section there are many titles including those on a variety of religious topics such as Magic, "New Age," or Witchcraft. Included, also, will be "versions" of the Bible such as the New International Version, the American Revised Edition, the Scofield Bible, or the Jerusalem Bible etc. Each of these will be available on the shelf ready to be sampled by the curious reader. However, copies of the King James Version are often not available or,if they are, they are wrapped in plastic coverings to prevent the casual browser from comparing this true Bible with the imitations flooding the market. This cannot be accidental. It represents a deliberate effort toward the suppression and eradication of the King James Bible by replacing it with bibles containing "another gospel." The effort to create these "other" bibles has been a massive undertaking. Imagine the financial cost of hiring armies of literary drones to rewrite the Bible verse by verse. Who but PAGAN would have the funds or the motive to underwrite such a project?

With a full understanding of the King James Bible's teachings, you will learn to employ the "Sword of the Lord" the full and complete doctrines of Christ. PAGAN'S prime method of defense is to infiltrate and to weaken each successive organized group as it arises in opposition. Once inside, it renders these movements harmless by taking authority over them and then leading that organization and its members out into an intellectual "wilderness." Therefore, the correct method of attack is never to "over-organize" or blindly place your full trust in any other person but to judge each one only by their fruits. Otherwise, act alone, and adhere to first principles. Trust only in the Word of God. Acting alone, mark out one person who seems receptive to hearing the Word and, who if he hears, must have been "called." Commence then, and expose the faith to him. Gradually teach him; teach him all you know. Arm him with the doctrines that will make him strong. Convert him to be a teacher. Set him, then, on his way to teach another. Do this with the next person as the opportunity arises. Include the agents of PAGAN if they will listen. These consider themselves to be liberal, progressive and logically minded but they, like everyone else, are seekers. They, too, are "true-believers' who, unfortunately, just believe the wrong things. They are the "hollow-men" (and women) as described by T. S. Eliot in his poem "The Wasteland"; "Hollow-men" who can be filled and restored by the Gospel.

We have a great biblical example. On the road to Damascus, Christ, at one stroke, converted his greatest enemy, Saul, the prosecuting attorney of the Jewish Pharisees, into His greatest Apostle. We may do likewise. We may begin by converting others one by one. In so doing we will be creating the world's greatest multi-level marketing system. If you and I begin by converting only two others what will be the result? If one of us converts two persons each of whom goes on to convert two others, by the time our multi-level pyramids become ten levels deep, 2048 new converts will have been made. By the twentieth level the number becomes 2,097,152!

During a public interview, the courageous Romanian underground Pastor, Richard Wurmbrand, (who, himself, was a converted Jew) was asked how Christians like himself survived under 50 years of Communist persecution? He said that since the penalty for exposure meant imprisonment or death, those who accepted Christ made a fervent commitment. So much so, that every convert became another missionary. When told by some concerned American patriots that they feared the arrival into this country of one hundred Communist consular officials, he replied; "Since you believe you are a Christian country, what are you afraid of?—convert them!"

A good example of how to proceed comes from the Christians of modern-day China. There they have adopted the "Home Church" method. While the Chinese Marxist rulers permit some Christian missions and churches to exist these must only operate under state supervision. Real Chinese Christians, knowing the official churches are infiltrated by informers, have set up a "home church" system by which they hold services and teach the Bible in their private houses in small groups. The movement is spreading like wildfire in China and is frustrating the efforts of the government to control religion. It is a brilliant method of organization and can be used anywhere.

Under present conditions it is absolutely necessary this work of conversion be carried on because the alternative is the victory of PAGAN. PAGAN seriously intends to regiment all humanity into one herd of domesticated animals placing us into what will become a grand socialist experimental farm. Make no mistake; this *is* their determined goal. They call their coming Utopia the "New Age" but in actuality it will be the return to the "Old Age" the "Old Order" of world wide Paganism: a "Pax Diabolus."

It is the duty of all Christians to ensure this does not happen. We must be the preserving salt of society that renders impotent the diabolical leaven of PAGAN. Our potential allies are everywhere. Around the globe, the common people have God's Word written on their hearts; whether or not nominally they follow other faiths. The common people instinctively seek the will of God. They long for order and for the opportunity to honestly prosper. Usually they are led astray. Their inherited religious systems based upon wisdom teachings and on semi-divine gods and spirits do not supply them with the guidance for living in peace. Rather they create only justifications for continuing conflicts with their neighbors.

Spengler's analysis of Western Culture describes this time as being the stage of "Civilization" wherein the culture is "Become;" having fulfilled its potentialities; able to replicate and to extend itself for a time but lacking any further potential for new intellectual development. The mark of this stage is universal skepticism.

Skepticism is seen on every side. It is the lack of moral commitment that makes thought-control possible. Today, it is prohibited to speak in public any thought not determined by PAGAN to be "politically correct." Those speech restrictions are accepted because the people have no heart felt convictions to defend. "They are going along to get along."

A worldwide a border-less intellectual politico-economic system has been set up which, while it is outwardly Americanizing the world, is at the same time causing a cultural homogenization; it is replication without any accompanying belief system. If continued, the final result will be the iron-rule of an all-encompassing godless tyranny.

There is a means of escape. While Spengler's analysis of the birth maturity and death of cultures is unassailable, his former subjects of study were cultures that did not understand their position on the time-scale and were incapable of self-regeneration. As the commitment to its moral traditions waned the Republic of Rome ended and the Empire replaced it. This First Roman Empire ran down in its turn and froze into an administrative gridlock that prevented any further regeneration. In about the year 500 AD, the peoples of France Spain and Germany neither believed in the Roman Empire nor took it very seriously. Can we avoid that ending? Can a better understanding of historical processes provide us with better guidance? Can we halt and reverse what will be an inevitable slide downward toward Imperialism? Yes, it is possible. It is entirely possible that this can be accomplished.

Because the fundamental basis of culture is religious, the answer for our renewal is a religious sea-change. We must create what Spengler predicted was a possible alternative to the end of a Civilization. He called this the "Second-Religiousness" wherein a culture might, by making a new religious commitment, found a *new* culture. The central problems of Western Culture are the questionings posed by our skeptical world-view. Skepticism, if not answered, will cause the demise of Western Culture; the "Decline of the West; just as Spengler predicted in 1918.

PAGAN recognizes the importance of religion. That is why they advance so many pseudo-religious schemes to entice the unwary. Each promises to bring about world peace; but only the Gospel of Christ has the power to produce a real and lasting peace.

We see the hand of PAGAN in each of the world's trouble spots. They say they work for "Peace" but there is no peace. In Ireland, Israel, the Balkans and elsewhere whenever the opposing sides in these civil wars attempt to resolve their differences, the agents of PAGAN throw more gasoline on the fires and peace is averted once more. These are the planned scenarios that are being repeated over and over. The news reports of the peace conferences and peace keeping missions of today could well be copied from the newspapers of 10 20 or 40 years ago.

THE BALKANS

To Westerners, and especially to Americans, the situation in the Balkans seems far away and impossible to understand. The conflicts between these small national groups, to us, seem irrational and unnecessary. What are they fighting about anyway? Why can't they simply get along?

The reason they cannot get along and why the conflicts continue is that their differences are not political; they are religious. With that understood everything becomes clearer. The religious and thus the ethnic and political makeup of these peoples are the result of nearly 2000 years of history during which three major religious forces have contended for control of the region.

On the map Yugoslavia can be seen to comprise a rectangular area along the Eastern side of the Adriatic Sea with Austria and Italy to the North and Northwest and with Greece and Turkey on the South. In the Northwest are the Croats who are Roman Catholic and are allied with the interests of the Vatican and Austria. In the South are the Bosnians and the Kosovars who, because their territories were occupied for nearly five hundred years by the brutal Ottoman Empire, in self-defense they converted to Islam. And, finally, in the center of the country and along its Eastern border are the Serbians who are Eastern (Greek) Orthodox and who traditionally are aligned with cultural influences from the Pan-Slavic Russians.

During the last ten years, what has happened in the Balkans has been an assault upon the Eastern Orthodox Serbians by the United States, in the religious and political interests of the Vatican. Secretly, the United States has armed and underwritten the revolt by the Muslim Bosnians (Kosovars) and the Albanians against the central government of Yugoslavia; controlled by the Serbian majority.

The Serbs are trying to somehow hold the country of Yugoslavia together but it is the policy of the Vatican, and thus of the United States as well, to dismember it and turn it over finally to the Catholic Croats by using the simple-minded Muslims as cannon-fodder. Fundamentally, the goal is to drive back or to overwhelm the power of the Serbians and their Eastern Orthodox Church in Yugoslavia and, by so doing, to advance the power of the New Roman Empire one click further to the East. (The determined goal of the Vatican for the past 1000 years)

RWANDA

The recent genocide in the African state of Rwanda is another example of the Vatican's duplicity. The underlying political fact in Rwanda is that it is composed of two opposing factions: the Tutsi and the Hutu tribes. The Tutsi known in the 19th Century as the Hottentots or as the "Totsies" by the Dutch of South Africa were historically the more educated and militant faction and the natural rulers of the country; even though the Hutus are numerically in the majority.

When, in the 1950s, independence from colonial rule occurred and the democratic principal of one-man one-vote was imposed the Tutsies voluntarily submitted to a new government formed and dominated by the more numerous Hutus. A fact never mentioned in the Western press is that the Tutsi tribes had been Christianized by Protestant missionaries; while the Hutus had been proselytized by the efforts of the Roman Catholic Empire. Thus, the fundamental tribal differences between these two groups were intensified by their religious differences.

After the death of the Hutu Rwandan President in an airplane crash blamed upon Tutsi sabotage, the government of Rwanda embarked upon an official plan to exterminate the Tutsi tribes entirely. For weeks, the Hutu people were exhorted by press and radio to prepare themselves for a "cleansing" of the nation. The Hutus were armed by the government with thousands of newly imported machetes. The Catholic Archbishop of Rwanda and his local priests actively joined in this plan because it meant the extermination of Protestants. The Tutsies knowing what was coming, appealed to the United Nations and to the Western powers for protection but mysteriously none of these potential rescuers seems to have heard them.

When the blow fell, Hutu mobs incited into fury by a combination of natural tribal animosities and by their Catholic priests, fell upon the Tutsies and massacred over 800,000 of them before their blood lust was sated. In response the remaining Tutsies formed themselves into an ad-hoc vigilante army which defeated the Hutu assassins and drove them out of the country. The Tutsies have now reasserted their political control and today it is the Hutu leaders along with the Rwandan Catholic Archbishop and his priests who are standing trial in Rwanda for the crime of genocide.

During the months-long period of this genocide, President Clinton was asked why the U. S. did not intervene to stop the massacre since we had intervened in so many other outbreaks of lawlessness in Africa? To this, he answered; "We cannot be everywhere!" Indeed one can understand why President Clinton who, lives under the direct control of the Vatican, would have no interest, whatever, in halting a Holocaust against Protestants in Rwanda or, for that matter, anywhere else

In Ireland, Africa, or Palestine, nothing has changed. The real news is there is no news. Under PAGAN there are only the repeated scenarios of agitation, propaganda, and massacres, designed to incite new hatreds. The purpose of PAGAN is to alienate all peoples races and religions one from the other; to maintain them in a condition of permanent war against each other; wars designed to destroy all cultures so that, finally, only the Second Roman Empire will be left standing.

Understanding PAGAN'S techniques but also knowing that, while it is powerful, it is rigidly committed to this single-minded method of operation, the answer to the world-problem becomes obvious. The answer is the Christian Message.

PAGAN depends for its very existence upon maintaining an atheistic environment within which all manner of sin and evil is promoted. This is an environment designed to further its program of social disorganization. Thus within the

United States and elsewhere under the interpretation of the civil laws according to PAGAN:

> Human rights are superior to property rights.
> Murder: Is justified if one feels socially disadvantaged.
> Robbery: Is justified if the thief has "needs"
> Infanticide: is justified if the mother is inconvenienced.
> Fornication: Is justified; "If it feels good do it."
> Homosexuality: Is justified; "Let's Celebrate Diversity."
> Blasphemy is justified' It is only "free speech."

The promotion of these sins against the morale of society constitutes internal acts of war by PAGAN against the civil population. Emerging nations say they do not so much fear the American military as they do receiving our corrupting MTV programs by satellite.

Just as it agitates to create wars between nations, PAGAN works within societies to promote every wicked and sinful practice imaginable. This is done to debase our citizenry so as to render them unable resist the chains of slavery. Brain washing is the method of PAGAN. It is its major weapon but it is also its Achilles Heel. To destroy it we must only reply with the brain-restoring message of the Gospel; refuting their evil with good; using the Word of God that provides the all-sufficient answers to a world instinctively seeking true peace.

The advance of communication technologies soon will require a single computer operating language. So, too, will the need arise for a world-wide system of belief. This must come; and this will be either the atheistic religions of PAGAN or the Gospel of Jesus Christ.

Christians must re-form their Church-Militant and mentally re-arm for the battle. As set forth above the Christian method of warfare must be the same as PAGAN'S; mind-conversion. Christians must study the Word so as to put on the full armor of God; commencing the process of converting others, one by one, and one by one, liberating them from the spells of PAGAN. This work of conversion will always be opposed because it strikes fear into the heart of PAGAN. But victory must be our purpose and so much as it is given for us to accomplish; to create the City of God upon this Earth.

The alternatives are clear. In the near future the Earth and its peoples will live in an earthly Heaven or an earthly Hell. As with the Israelites in the time of the prophet Samuel the day has come when we must choose—either to stand on the side of the living God or bend our knees to Baal.

Time Scale. AD 100 = AD 2000

CHAPTER ELEVEN

CHRISTIAN CULTURE

Two religious forces contend for the world today. One is the Culture of PAGAN and the other its opposite Christian Culture.

In large part Spengler's "The Decline of the West" describes the desperate rearguard struggle by PAGAN to regain its power over humanity. The Jesuitical Second Roman Empire and its allies are fighting to avoid their inevitable fate. In opposition, Christian Culture has grown from its springtime into full maturity. Consider these realities:

After incubating for a thousand years under the Inquisitions of the Middle Ages, Primitive Christianity and Christian Culture burst forth as if, new-born, in the 16th Century during the Protestant Reformation. The result of this new birth, (The Renaissance), was an explosion of intellectual and technical development that has created the modern world.

Liberated from the mental and social straight jacket of the Second Roman Empire, the talents of the common people were suddenly released in a flood of innovation. What began then, and continues on, to the present, is a whirlwind of social and technological progress. Modern "Progress" is the product of the Protestant Reformation.

> Because the Reformation was truly Christian it observed "Fair Play."
> Each person gained the right to own real property.
> All became equal before the law.
> Citizens gained the right to make private contracts and to have them recognized at law.
> Individuals gained the right to profit from their labor.
> Business "corporations" were created; permitting large groups of free individuals to joint venture to create larger enterprises.
> Intellectual property was protected. Inventors were permitted to profit from their inventions.
> Christian standards of truth and honesty in financial dealings became the legally accepted norm.
> Sound money was fostered; making economic transactions reliable.

Christian governments responded to the needs of the people.
The Protestant Reformation founded modern Western civilization and
the free enterprise economic system.

Many believe that "Progress" has been the natural upward march of mankind over thousands of years. But this is not true. "Progress" was unknown under the rule of PAGAN. From pre-historic times, down to the 16th Century AD, there was little improvement, whatsoever, in the living conditions of the common people. For them, life was lived at the subsistence level. As of the year 1500 AD, travel was still by foot or on horseback over roads ankle deep in dust or mud. There was nothing but wood for fuel. No one had discovered a use for coal or petroleum. There was no lighting at night except by candles; no machines to take the weight of manual labor from off the backs of the peasants. There were no real doctors or any useful understanding of medical science. Plagues decimated the population every few years without anyone understanding the cause. If sick, one simply lived or died

Prior to the Reformation, the Second Roman Empire and its feudal system frowned upon new ideas as dangerous. Progress was discouraged by the decree that all of the land and the common people on it, and their possessions as well, "belonged" to their lords. The peasants were only sharecroppers. Routinely, anything of value held by the peasants would be seized by their rulers in the form of rents or taxation and wasted by them in wars or upon vain luxuries. The common people were chattels; having no more legal rights than pigs or oxen. The Catholic Church was allied with the aristocracy and, together, they kept the peasants in poverty and in ignorance. Commoners were not even given an opportunity to learn to read or write! This is the same social system that PAGAN intends to re-impose upon mankind under the cloak of its One World Religion and its New (Old) World Dis-Order!

In contrast, under Christian Culture, what wonders have occurred? Within one hundred years after Martin Luther nailed his 95 theses on the door of Wittenberg Cathedral, Protestant Dutch, and English, private corporations were founding colonies of free Christian, property holding, men and women in the New World. Within another hundred years, coal and steam power came into use. In the following hundred years, the Industrial Revolution began and soon came the invention of the steamboat, the railway and the telegraph. The following Century saw thousands of useful inventions including the miracle of electric power, the telephone, motion pictures and most of the basic equipment we now use in modern life. The 20th Century produced a staggering number of further advancements including the automobile, radio, television, air travel, and the computer.

These advancements were created, almost entirely, within the Protestant Christian countries of Europe and North America; not elsewhere. During the 19th Century, in Great Britain, while the Industrial Revolution gradually raised the standard of living, the Catholic peasants in Ireland, but two hundred miles to the West, were still living in stone huts and subsisting on a diet exclusively of potatoes. When the blight destroyed this single crop, one-third of the people of Ireland died of starvation. England was criticized for not providing more relief during the famine, which may be true, but the basic cause of the famine was the primitive conditions in Ireland fostered by their Catholic Church. In fact, when their English set up soup kitchens to feed the starving people, the priests told them that if they partook of this damnable "Protestant Soup" they would go straight to Hell.

Throughout the world, the root cause of poverty is always religious. The starving beggars on the streets of Calcutta are but the end product of the Hindu religion that determines the course of Indian culture.

Comparing the old Catholic-feudal economy with the new Christian-Individualist system offers a contrast in opposites. Not so very long ago, in Cremona, Italy, during the period from 1675 to 1730, the Stradavari family made some of the finest violins ever produced. Many are still in existence and are highly prized by famous musicians. But the secret of *how* they were made is a mystery. Whatever their special secret was, the Stradavarii were forced to hold it closely within their family because, under the feudal system, if they had published their secret it would have been copied by others and no longer be of value to them. Under feudalism, they had no legal right to their discovery. The logical course, for them, was to limit themselves to making only the number of instruments they could produce by hand in their little shop. But when the last of the family died; the secret died with them.

Imagine, however, that the famous modern American inventor, Professor Guido Stradavari, lived in our present day Christian-Individualist culture. As soon as he realized he had discovered a process for making superior violins, he would immediately publish this information by applying to the U. S. Patent Office for a patent. Once it was filed, he would then organize a factory to make his super-violins in volume. He might, also, license the process to other violin makers; charging them a royalty fee. In this manner, he would be immediately compensated. Then, after 17 years, when his patent had expired, his invention would pass over into the public domain and become a permanent addition to the world's store of knowledge. This would not be a great loss to the famous inventor because, by that time, with the profits from his first invention, he would no doubt have even more important inventions under development in his modern research laboratory.

Modern technologies are the product of an accumulation of separate inventions that are now in the public domain, and not simply the result of a single person's genius. The computer is not a single invention. It is an assemblage of thousands of separate inventions that, bit by bit, contribute to the final product. This process is not static; as newer and better ideas are found they are added to the products, year by year.

The pace of modern innovation is blindingly rapid. The transistor was invented only 50 years ago. The first laser beam, now an essential component of the read/write compact disk system, and of fiber-optic communications, is only 40 years old. The PC computer revolution started only 25 years ago! With such examples before us the great political question becomes; shall this wondrous cornucopia of Christian-Individualist culture be halted, lost and forgotten, by our returning to the Imperial Roman Catholic-Feudal culture of PAGAN?

While the failures of Socialist economies are well recognized, the connection of religion with these failures is generally overlooked. In socialist countries, because their religion is officially atheism, their planned economies must be operated by force; from the top down. This is why they fail. In every country where centralized planning is attempted the result is always the same; it is a disaster. As it was in the Middle Ages there is never enough of anything; rationing or starvation is the rule. Socialism does not make people equally rich; it makes them equally poor.

In contrast, under Christian Free Enterprise not only is there enough of everything; the economy produces in such abundance that "overproduction" is the only economic problem.

In a recent book, entitled "The Mystery of Capital," a brilliant Peruvian economist, Hernando De Soto, gives us an explanation of just *why* capitalism has "worked" in the "advanced" nations and *why* it has failed to take hold in the "third world." He found that, in the third world, in cities like Lima, Peru or Cairo, Egypt, or in cities elsewhere, the common people still live under the rule of feudalism. While they may be paid wages, as workers in the city, or as sharecroppers on nearby farms, the people do not own title to the home they live in or to the farm they cultivate. If they build a house on some open site that no one else is using, they remain squatters because they will never have title. Thus, even though they may be permitted live there, and their house may be paid for, they have no way, legally, to sell it to anyone else or the ability to obtain a mortgage on it to liberate their capital. The fundamental fact is that, in these countries, the capital of the whole nation is "locked-up" by their feudalistic economic system. This is why they are "poor."

In individualistic countries (i.e. those based upon a Protestant legal tradition) Germany, England, Canada and the United States, a fact, totally unnoticed by their own economists, is that their powerful economies were founded upon their always having had a local Registrar of Deeds Office where, in every district or County, titles to all real estate properties are recorded and kept. Those deeds are supported by the justice system that defends them. This means that, in these countries, the common people can own real property, and that buyers, sellers and financial institutions can rely upon the validity of their deeds. It means, further, that the capital locked-up in this real-estate may be liberated by borrowing on it with mortgages creating economic liquidity. The resulting funds are then used by borrowers to finance other enterprises. The mortgages generate "interest" for the lenders who then re-loan their profits, at a somewhat higher rate, to other business entrepreneurs.

For the government of any feudal country, converting to a private property system would have an important benefit because, when a citizen receives a registered deed to real property, he is forced to come out from the "black economy." The government now knows who he is and what he does. In this way, he becomes a "taxpayer" and if he fails to pay his taxes the title to his property can be held hostage.

In Protestant nations the development of this system seemed so natural that people neither recognized its enormous value nor remarked upon it. It took a clear-eyed scholar from the third world to recognize that it was the system of private land ownership that creates capital wealth in the advanced countries. The lessons taught by this book, "The Secret of Capital," are now being studied by the governments of several under-developed nations. If these find the political will to apply this "secret," they will soon be able to raise themselves from poverty into prosperity.

Protestant Christianity is the world's most successful culture. Its productive capacity continually defies all the attempts of PAGAN to contain it; even with brutal rates of taxation and intensive government regulation. Non-Christian skeptics, who may be satisfied with their present economic condition in America or in Western Europe, fail to understand that their good fortune is only made possible by their living within a Protestant economic system. Let them try to succeed in a non-Protestant country where private property, "fair-play," and honest dealing, are the exception and not the rule. Socially, what the Protestant Reformation created was something new in human history. By its recognition of the sanctity of private property, it permitted the development of an economic third class; the "Middle Class." This had never happened before. Prior to the Protestant Reformation, there had only been two economic classes; the Rich and the Poor.

Commencing with the Renaissance, the expansion of learning, resulting from the invention of the printing press, began to threaten the iron rule of the Church.

The Pope and his Bishops were well aware of the danger and did all in their power to suppress new discoveries because they deemed them to be subversive. New ideas were condemned even while the Popes, the Cardinals, and their Bishops knew them to be true and useful. This was done because those new discoveries threatened the psycho-political reign of the Second Roman Empire.

The unyielding rigidity of the Church is demonstrated by the trial of the astronomer Galileo, in 1633, to suppress his discovery that the Earth orbited about the Sun. Galileo is still on trial. In the eyes of the Church he was a heretic then and is a heretic now.

While the Protestant-Christian Reformation did not officially begin until 1517 it was preceded by the Albegensians, the Paulicans, the Lollards, the Hussites and by all the Primitive Christians of Europe who had, for a thousand years, maintained their faith through the terrible persecutions rained down upon them by the armies of Rome. As the Protestant rulers broke away from Rome, they organized Christian governments in England, Germany, and in Scandinavia. To re-conquer those reformed nations, the Second Roman Empire, and its Jesuits organized a multi-national War of Counter-Reformation against the Protestant States.

That War of the Counter-Reformation continues on today. It was and still is a reactionary movement; an attempt to rein in and to control Christian Culture. But because of its need to operate by clandestine methods, the Second Roman Empire has failed to achieve its goals. In the Western world, Christians *are* still the "Silent Majority" It is only because they have not responded to the challenge, that PAGAN remains able to remain in political control.

THE CHARACTER OF PAGAN

The character of PAGAN is symbolized by the pyramid seen in the Great Seal of the United States and as displayed on the U. S. one dollar Federal Reserve Note. That pyramid symbolizes the program of PAGAN which is to "centralize" all power into one pyramidal system of "command and control." Socialism is the political and economic system mandated by the godless religion of PAGAN. Roman Catholicism *is* Socialism! Socialism is Roman Catholicism! Roman Catholicism is the fundamental philosophy behind all socialist parties and regimes in the world. There is no philosophical difference between Communism, Nazism, the Papacy or the "New Deal." Each demands that all decisions originate at the highest level. Plans are then formulated and orders given; and whether they come from a Czar, a Pope, or a Fuhrer, makes little difference. Everyone below is expected to obey.

We have all seen the attempts to impose both national and international varieties of "socialism" upon many nations but what is pure socialism; itself? The answer is that pure socialism is the social blueprint of the Jesuits and of the Catholic Church. "Socialism" is the Vatican's codeword for its plan to thrust its age-old system of feudalism once more down upon the peoples of the world. Under this program, private property will gradually be taken back (viz. Cuba, Zimbabwe and Venezuela) and owned by the state or by the Church. As in the middle ages society will be divided into two classes. The Poor, the lower three quarters, who are the productive class who will work virtually as share-croppers or as wage-slaves, living in public housing and keeping only enough live on after being brutally taxed to support the non-productive upper one-quarter—the Rich. In this upper one-quarter, the Rich and non-productive class, there will be the popes, the bishops, the priests, the monks, the nuns, the administrators, the educators, the politicians, their staffs, the legal system, the penal systems, the presidents, the kings, the commissars, the dukes, the lords, the famous actors, and, of course, the military caste necessary to maintain order and to put down the occasional revolt.

Those who have served in the military will find this system familiar. They will also understand its fundamental weakness. While the Commander-in-Chief may decide to seek a battle, by the time his orders have reached the front lines, at the bottom of the pyramid, the soldiers may be far less eager than he is to fight the battle and die.

The inefficiencies of socialism can be seen in every governmental agency. Under socialism, while the individual may be willing to give his best, the system is against him. He finds that whether he is a hard worker or a slacker the pay is the same. He is told to follow orders and that, if he receives no orders, he is to do nothing. Acting without orders is an offense to his superiors and liable to be punished. Finally, he learns to "play the game" which is; to do as little as possible, not to make waves and to patiently wait for his pension.

This may be an acceptable compromise for the individual but he now represents a drag upon the economy because he is costing more (and consuming more) than he produces. Magnified fifty million times as in Soviet Russia it brings the economy to a halt.

The fundamental weakness of Pyramidal Socialism is that, even when a plan is correct, its effectiveness is diluted as the orders pass down the chain of command. Good plans produce only mediocre results. On the other hand, should the plan contain some fundamental error, that error becomes magnified as it trickles down to the lowest level. That original error now produces a disastrous result. That is what is wrong with Russia.

THE RUSSIAN DILEMMA

After having, technically, changed their economy from the communist system to what they believe to be free enterprise, there still has been little real improvement in the Russian economy or in the lives of its people. The International Monetary Fund has poured billions of dollars into Russia to prime the economic pump but it has all been in vain. Whenever any real cash appears in the accounts of a business, it is immediately siphoned off by whomever gets their hands on it first and it is then transferred out of the country into the private accounts, of the embezzlers in some Protestant-Capitalist nation This wholesale dishonesty continues because, as economic realists, the embezzlers recognize that Russia has only exchanged Soviet Socialism for another form of Socialism and that, otherwise, little has changed. One Russian pessimist has been quoted as saying "There is no hope for Russia. We have always lived in the mud and we will die in the mud; nothing will ever change"!

Why is this so? The answer is again religious. Since the conversion of Russia to the Greek Orthodox Church (A Catholic variant) in Kiev in 988 AD, Russia has always been a feudal socialist state. Feudalism is all the people know. They cannot make free-enterprise work because they do not understand it. For the last 1000 years they have been ruled over as subject peasants of the Czars and of the Orthodox (Catholic) Church. Even when the last of the Czars and the Church were overthrown in 1917, there was no outbreak of personal freedom. The people were not ready for freedom. Instead it was found necessary to set up an even more cruel form of feudalism under a series of new "czars," Lenin and Joseph Stalin, et al., and even today, to a degree, under Chairman Putin. Since 1989, in cooperation with the West, the socialistic Russian politicians have attempted to set up a free economy but they find themselves powerless to make the necessary fundamental changes. Underneath it all, the Russian people still hope that someone will take charge and reestablish a workable economic order. By this they show that they are still, at heart, slaves looking for a kindly master!

What they still have in Russia is a pyramidal "top-down" government along with a pyramidal "top down" economic system. The people are used to this and, being highly conservative, are reluctant to change. The governors likewise know that the present system somehow works even though at an inefficient level.

What they cannot see; we can. That is: the answer to the Russians' problems is that before they can prosper, economically, they first must become Protestant Christians. Protestantism will give them the fundamental belief in their own personal and legal individuality. When they do this, they will gain the right to the ownership of real property. When this conversion might take place is problematic; for they do not have behind them the centuries of experience in respecting the rights of private property that were formalized by the Protestant Reformation in England, Holland, Germany, and America. Economic freedom demands first that the Russians throw off the Orthodox (Eastern) Catholic Church and accept their individual salvation by Jesus Christ and adopt as corollaries respect for the individual's right to valid deeds in land and in all the other forms of private property; and to establish a government which will defend those rights. It demands a universal acceptance of the idea of a justice system that can be trusted because it is religiously based and thus impartial.

The only answer for Russia is a religious reformation. This means a fundamental conversion of their minds and hearts into a new way of thinking. This must be on the part of all involved; on the part of the leaders as well as of the people. When this happens, and only then, will Russia develop into a truly modern country.

JAPAN

A useful lesson can be learned from the forced conversion of Japan from their old system of feudalism to their new system of free private enterprise. After the end of the Second World War, General MacArthur and his military government, set out to impose upon the Japanese, our concept of the private ownership of property.

All across Japan, the people who were farming their plots of land were dependent upon tradition to protect their personal rights. They had no actual proof that the land was theirs because they had no deeds. What the Americans did was to make a Land Survey of Japan and then they presented to every present owner who could prove he was the valid proprietor, a registered property deed. This fundamental change underlies the economic reform of Japan, from feudalism, into a modern nation. Might this not be a valid first step for the Russians to copy?

THE CULTURE OF CHRISTIANITY

The opposite of Pyramidal Socialism is Christian Culture. Under Christianity all is reversed. Here the individuals, at the bottom of the pyramid, the consumers, are theoretically supreme. Power and decisions rise up from below. Political power is assumed to rest in the people, at large. These decide policy; delivering their commands to their selected leaders. Control is from the bottom-up; not from the top-down. This is the theoretical basis of the United States' representative form of government; and, were it honestly put into practice, it would create a Heaven on Earth.

Business under Christian Culture produces more at less cost. The remainder is profit. Profit is the exact opposite of the disincentives of socialism; wherein any production beyond the plan is suspect and goes unrewarded.

The bane of PAGAN'S existence is that Christian freedom has created so vast and so successful a society it appears to them nearly uncontrollable. Regulatory agencies must constantly be created to monitor the complex activities in progress. Programs of brutal taxation must be devised to sap away the strength of the Christian economy. In truth the key problem for the governments of PAGAN, is not in finding enough money to operate, theirs is the problem of finding ways to increase waste so as to dissipate the excess funds constantly being supplied by the punitive tax system. Those taxes are imposed not for revenue but for control.

We are involved in a war that began in 1517 with the rise of the Reformation: "THE WAR OF THE COUNTER-REFORMATION." This war continues on today. PAGAN'S main target, as in all socialist regimes, is "the Bourgeoisie" or "The Middle Class;" because they know that, while the peasants and workers can be bought or controlled, and that Big Businesses can be taken over and run by the state, the Middle Class is a constant threat to their authority. The "Middle Class" is the first target of PAGAN.

PAGAN'S top-down method of subversion is to infiltrate the centers of intellectual, political and financial power. Their members seek positions of influence in the universities, in politics and in the media. Then, as they gain in numbers they promote one another to higher and higher levels. Politically, they make every effort to control the electoral process from the district party caucuses on up. In this way they control the selection of approved candidates in all the major parties. At times, they support one candidate over another but whatever the result of an election their candidate always win. They dominate international finance including the foreign-owned Federal Reserve Bank. This permits them to counterfeit paper "money" at will with which they hire armies of agents to perform their bidding.

Despite its apparent strength, PAGAN is still a parasite. It produces nothing of value and feeds from the body of the only productive host it can find, Christian Culture. It milks this Culture with taxes, robs it again with printing-press money, cripples it with regulation, and binds it in a web of punitive laws. But without the compliance of its host PAGAN would wither and die.

The awakening of this host is the constant fear of PAGAN because when, Christian Culture learns the true nature of PAGAN; it will throw it off; rising to assume its rightful place. PAGAN understands this danger and makes every possible effort to silence all sources of information that might advance the Gospel of Christ. Television evangelism is repressed, denominational leaders are cowed by threats to their tax exemptions—the print and TV media ridicule Christianity and Christian morality as "old fashioned;" while the King James Bible is rewritten to dilute its message. This is done because PAGAN well understands, if Christians do not, that the struggle for world dominion is a religious war and nothing else.

Christian Culture holds the key. It has the power and the God-given right to throw off the spells of PAGAN and to reign supreme as the dominant and saving culture of the world.

CHAPTER TWELVE

THE CHRISTIAN RESPONSE

The answer to PAGAN *is* Christian Culture. Christians must take up their proper roles in opposition to PAGAN; committing themselves to the fundamental articles of the Christian faith. Renewal will create a new "Church-Militant" which must be established before we can reaffirm Christian Culture. But before this can happen, there must come an awakening by individual Christians to the fact that they are in mortal danger; that there *is* an enemy that intends to destroy them both mentally and physically.

The main-line denominations smugly believe that the battle for their liberty of conscience was won for them during the Reformation by the establishment of the Protestant Churches and by the deaths of millions of Christian martyrs when, in fact, nothing conclusive was won at all. The Reformation bought only a truce, not a final victory. Following the Reformation, their arch-enemy, the Second Roman Empire, regrouped and began a new war; an underground war of subversive Counter-Reformation. The Second Roman Empire never surrendered. Its aim remains what it always has been; A One World, Earthly Dominion.

While each believes it represents the one true faith the Protestant denominations and the semi-Christian cults are being led about to follow after every imaginable, unbiblical, perversion manufactured for them by the lies of PAGAN. Doctrines have been introduced to neutralize the power of Christianity; distortions such as Premillenialism, Arianism, Pentacostalism, Unitarianism and Evolution; all of which are founded upon the doctrines of mere men to the near exclusion of the Word of God. These heresies have so undermined their moral foundations that the Protestant denominations now accept doctrines and behavior that once they would have found abhorrent. All is now justified under the relativistic blanket-excuse of; "God's Universal Love."

Adrift without any heartfelt beliefs these Protestant denominations sense the gravitational pull of the massive and unyielding Second Roman (Catholic) Empire. This leads them ever-downward toward the maelstrom of Ecumenism.

Unaware of their danger, the Protestant clergy wax fat in their sloth; living lives of comfort and ease upon the tithes and gifts of the poor. Unctuously, they carry out their duties by rote while their flocks go unfed; never hearing from their

ministers the fundamental doctrines of the Christian faith. These are truly a generation of vipers who make merchandise of the Gospel. By their acts of commission and by their acts of omission they sell their souls to PAGAN.

This must all be cleared away by a new Church-Militant. This New Church
must be created by the people. Created from below; created by the Christian laity
who can use the religious liberty granted to them by Christ. The laity has the
obligation the right and the power to overthrow their clergy and to set up a new
form of independent church government.

THE CHURCH-MILITANT

Already, many true Christians are withdrawing from their soul-dead denominations that have abandoned them. The Church-Militant will welcome into itself
these and all who wish to follow the Gospel of Christ; Catholics most of all.
These must be brought to see that they must follow the full Gospel of Christ to
gain their religious liberty. They must join the Church-Militant and serve no
longer as slaves of Paganism. They should be reminded that, historically, the
whole Protestant movement was founded by reforming Catholics who "protested'
and who repudiated the un-Christian basis of the Second Roman (Catholic)
Empire. By joining the Kingdom of God, Catholics will not be abandoning their
faith but regaining it.

In the Church-Militant, agreement upon one Bible must be fundamental. The
Old and New Testaments were set forth without blemish in the 1611 King James
Translation into English. Without agreement upon one Bible, there can be no
agreement upon one doctrine. If by nothing else, the divine nature of the King
James Translation has been recognized by the massive efforts by PAGAN to
destroy it. For the past four centuries, they have made every effort to revise and
pervert its message. The first attempt was to create a new translation of the Latin
bible of Jerome for the Catholic Church. This work is called the Douai Bible and
is the authorized bible for Catholics in use today

Subsequently, millions of dollars have been spent paying literary hacks to create "scholarly" revisions of the King James Bible; attempting to improve upon the
Word of God! Examples are; the Scofield Bible, The Revised Standard Version,
The New International Version and a host of others; all devised to strip the Bible
of its poetry and of its message. Key words have been eliminated and replaced by
others to alter the text and to destroy the meaning of the Gospel. Here is a quote
from the preface to one such "corrected" Bible:

"The Authorized King James Version is used as the text of this Bible.
However it is *generally recognized* that this translation contains many

obsolete words and phrases that may not be easily understood by the reader. To *clarify* the meaning of these difficult words and phrases more than 5500 translations from the American Revised Version have been added [in brackets] to the text. This arrangement enables the reader to see at a glance the modern translation of the obsolete word or phrase." Upon examining the text, however, it was found that these 5500 "corrections" occur mainly in the Epistles of Paul. Obviously the aim was to weaken and distort the doctrinal teachings of this greatest of the Apostles.[22]

These bowdlerized versions are so unreadable as to deny the laity their rightful access to the Bible. These counterfeit bibles go far to explain why modern Christians have only a slight knowledge of the Bible's fundamental message.

CHURCH ORGANIZATION

Just as Christ was the first to proclaim equal rights for each believer; by derivation, His church must be organized in an egalitarian manner, not as an "established" hierarchical church.

The one who is "called" should preach and if others wish to hear him he shall be their minister. He should teach and his hearers should listen but no man should "own" another. The minister is not an employee of the congregation. He is the servant of God. If satisfied, the congregation should provide for his needs as it deems appropriate. Likewise, if dissatisfied, individuals in the congregation are not compelled to remain but have the right to withdraw and to seek another teacher.

Preaching should be based upon the Word of God as expressed in the Bible and this is a test that should continuously be applied to the minister by the congregation. Deacons or others should have no right to "instruct" their minister as to what to teach or emphasize. The minister has been "called" not them. The purpose of any lay committee is to handle practical matters and to support the work of their minister. In such manner individual churches will remain free from misunderstandings and divisions.

As a reminder to all; here are God's commandments whose authority and wisdom cannot be challenged:

---

22   Milton, Joyce. "First Partner" Hillary Rodham Clinton. William Morrow Co. New York, p.71.

THE TEN COMMANDMENTS
(Abbreviated)
For the full text see Exodus: 20

God spake all these words saying:

1- I am the Lord thy God. Thou shalt have no other gods before me.

2- Thou shalt not make unto thee any graven images. Thou shall not bow down thyself to them nor serve them.

3- Thou shalt not take the name of the Lord thy God in vain.

4- Remember the sabbath day to keep it holy.

5- Honour thy father and thy mother.

6- Thou shalt not kill.

7- Thou shalt not commit adultery.

8- Thou shalt not steal.

9- Thou shalt not bear false witness.

10- Thou shalt not covet. Thou shalt not covet thy neighbor's house thy neighbor's wife nor anything which is thy neighbor's.

These are God's rules for just living. Violate them at your peril!

FREEDOM FOR THE CHRISTIAN CHURCHS

A major weapon used by PAGAN to control religious teaching is the granting of the tax-exemption. The tax-exemption is commonly believed to be a governmental subsidy in support of religion but this is far from true. The purpose of the tax exemption, in the eyes of PAGAN, is to control the preaching of the ministers to within the bounds of what it deems to be politically correct."

The nature of this tax exemption is misunderstood. To receive the exemption most churches believe they are required by law to form a 501 c 3. non-profit Corporation; and most do. But their error is that they do not *need* to become a 501 c 3. corporation in order to become tax exempt. Churches are already tax-exempt. Religious organizations are automatically tax exempt under the constitution; without any permission by the IRS. Forming a (501 c 3.) non-profit corporation does not give a church any additional tax exemption, it only gives its donors the ability to treat their donations as charitable deductions on their personal IRS Form, 1040. The most wicked result of this (501 c 3.) trap is that the pastor of a church must thereby restrict his speech and his activities to the purposes stated in the corporate charter and, specifically, that he may not engage in any social or political comment. Otherwise, the tax-exempt status may be withdrawn.

Accepting corporate status is unnecessary. Churches have always been tax-exempt under the U. S. Constitution. The use of this natural exemption explains how the "reverends" of the left; such as Jesse Jackson, Al Sharpton, and Martin Luther King have always been free to engage in political and social activity without being questioned by the authorities. They are free to say and do as they please *precisely* because they have *not* filed to be a 501 c 3. non-profit Corporation.

Any church that operates under the (501 c 3.) corporate status is not free. It is a controlled arm of the state. For Christian churches; the dissolving of their 501 c 3. corporation, should be their first order of business. Following such a "declaration of independence," churches and their pastors will become free to enter into the body politic without restriction. Then, the arrogant assertion that there must be a legal separation between Church and State will immediately fall into the dust.

GUIDING PRINCIPLES

(These guidelines are derived from the program of Alcoholics Anonymous)
We admit we are powerless over sin—that our lives have become unmanageable.
We believe that only a Power greater than ourselves can restore us to sanity.
We make a decision to turn over our lives to God through His Son Jesus Christ.
We make a moral inventory of ourselves.
We confess to God the exact nature of our sins.
We humbly ask Him to remove these sins.
We agree to make amends to all we may have wronged.
We continue to take inventory and when we are wrong promptly admit it.
We seek through prayer and meditation to improve our conscious contact with God praying only for knowledge of His will for us and the power to carry that out.
By having a spiritual awakening as the result of these steps, to carry this message to others; and to practice these principles in all our affairs.
Churches should be autonomous except in matters affecting other groups.
Each church has but one purpose which is to carry its message to those who still suffer.
Every church should be self-supporting; <u>declining</u> outside contributions.

Preaching the Gospel of Christ, the Church-Militant can become the most powerful force in the world in opposition to PAGAN. The main strength of PAGAN lies in the secrecy by which it controls the minds of its slaves. This creates an army of cowards who live in the darkness and who cannot appear in the light of day. As in the story of the "Wizard of Oz," The Wizards of PAGAN work from behind a screen of deception; pulling on the levers of power. Exposed, it will be seen what it really is: a minority organization of mean-spirited weaklings who aspire to rule the world through lies and deceit. In the light of God's truth they will shrivel and die. The Sword of the Lord will cut them all to the quick.

Because religion determines culture, an awareness of the importance of an overriding system of religious belief, daily, becomes more evident. This sea change in the world's thinking already is underway. Around the world, people seek social solutions based upon spiritual foundations. The rise of Muslim fundamentalism is symptomatic of this fact. In Western societies, Christians and others grope toward the conclusion that a religious commitment must be superior to governments. Lay movements such as the Promise Keepers have shaken off the spells laid upon them by their liberal churches. They have leapfrogged over the heads of their soul-dead denominational priesthoods to return to the true faith—standing unreservedly for the Gospel of Christ. These movements represent a ground swell from below and are a warning to PAGAN that a society based upon an atheistic state of "social disorganization" is unacceptable. It will not work, and it will not for long be tolerated.

The Church-Militant will conquer by openly and fearlessly preaching the Gospel of Christ—period. There is no other way. It is fruitless to believe that political solutions are possible. As the work of conversion goes forward, politics and politicians will follow. The Gospel of Christ is the "Message." It is the message PAGAN cannot answer. Christ conquers all.

EPILOGUE

Today we may contact any person on the planet who has a computer and a telephone line. Soon we will be involved in one single worldwide conversation. The result will be the formation of a single world-view that will develop finally into a single organic religious view. It is the destiny of Christian Culture to be that world-view.

The charlatans of "Science" are being driven back into the shadows by the advances of true science. The ridiculous "Theory of Evolution" has become less and less defensible. God's truth, daily, is becoming more self-evident. Our Twenty-First Century science presents us with entirely new vistas of our world and of the universe than were imaginable in Darwin's Nineteenth Century. Consider the pictures of the Earth taken from inter-planetary space. These reveal our world as it really is; a blue and white ball floating utterly alone in infinity; accompanied by its life-giving sister-planet, the Moon.

More sobering yet, is a picture taken by the Voyager 2 spacecraft from the region of Saturn looking inward toward the Sun. Transmitted back to us, in this photograph, are two star-like specks of light; the brighter of which we are told is the planet Jupiter and the lesser light our Earth-Moon system. This astounding vision of our ultimate situation, within the Universe, makes mockery of the idea that we can judge God or second-guess Creation.

The cosmological theories of the Evolutionists are pure balderdash as is the anthropomorphic idea of the "Humanist" position that "Man is the Measure of All Things." How thousands of highly regarded and "tenured" professors and other educated fools could have swallowed these lies for two hundred and fifty years only supports the opinion of the late G. K. Chesterton who. once said, that "when one becomes an atheist he does not believe in "nothing"—he believes in anything!" The ultimate secrets of creation will never be discovered by living men. Scientific training does not make men co-creators. True science is the exploration of God's universe; but humility before God must still be the foundation stone of the scientific method.

In the microcosm, the fundamental nature of the forces at work; electro-magnetism, gravity and light remain inexplicable by human reason. While these are useful in practical applications, their inner nature remains a mystery. Electro-magnetism, gravity and light operate by "action-at-a-distance," a possibility heretofore undreamed of. Action-at-a-distance means that our perceived "real" world functions not by way of mechanical cause and effect but by a form of "knowledge;" by which every electron and atom recognizes its relationship with every other electron and atom in the universe. This scientific fact should force serious thinkers, literally, to their knees in a state of awe and wonder.

All on this planet belong, regularly, on their knees worshiping what is for us the unknowable wonder of life. Thus man and God may reconcile. There is no need for human conflict. In a state of wonder and in recognition of the mercy and providence of an all-powerful God will come that final peace.

Sonnet XIX On His Blindness
John Milton

> When I consider how my light is spent
> E're half my days in this dark world and wide
> And that one Talent which is death to hide
> Log'd with me useless though my Soul more bent
> To serve therewith my Maker and present
> My true account lest he returning chide.
> Doth God exact day-labour light deny'd

I fondly ask; But Patience to prevent
That murmur soon replies God doth not need
Either man's work or his own gifts who best
bear his milde yoak they serve him best his State
Is Kingly. Thousands at his bidding speed
And post o're Land and Ocean without rest:
They also serve who only stand and waite.

CHAPTER THIRTEEN

CURRENT EVENTS

In the war between PAGAN and true Christianity, developments continue to unfold that support our prior conclusions. These are noted and recorded as they occur.

Motivated by their true purpose which is the overthrow of Christian Culture, the Jesuit priests and their Maryknoll nuns are openly moving farther and farther to the Left. In South and Central America, the Jesuit priest's and Maryknoll Nun's political activities are carried on openly. The Sandanista-Marxist revolution in Nicaragua during the 1980s was led by Jesuit priests who acted, first as military advisors, and afterwards as cabinet ministers in the resulting Marxist government. To the masses of Latin America the Jesuits preach that the people are "poor" because of North American capitalist "oppression." They proclaim that there must be a holy war waged against the "capitalists" What they really mean is there must be a holy war against Protestant Christianity. The Jesuitical campaign to set up socialist governments in Latin America is only a rehearsal. Ultimately they mean to overthrow "Capitalistic Oppression—everywhere."

Curiously, the Jesuits' concern with the "poor" in Latin America does not extend to the plight of the people of Cuba. The "poor" of Cuba are passed over because the Jesuit's Revolution in Cuba has already been accomplished. Cuba was their first Latin American demonstration project. The hero of that revolution, Fidel Castro, is one of their own. A never-mentioned fact concerning the early life of Fidel is that, before he decided he was a Marxist, he was being educated at the Jesuit Seminary at Santiago de Cuba. With his Jesuitical background exposed, the mystery of why the Castro Revolution succeeded so easily and who supported it becomes clearer; and it explains why the CIA supplied him with the necessary arms and money. It explains, also, why the Liberals in the United States have, for so long, tolerated Castro and left him in power. Be forewarned; the feudalistic tyranny now in control of Cuba is the same system that the Vatican intends to impose upon the rest of the world. Indeed as a further confirmation of Castro's real identity, Pope John Paul II, himself, made a pilgrimage to Cuba in 1999. When he arrived, he was warmly welcomed by his spiritual son and secret political agent, Brother Fidel, SJ.

On December 20, 2001 the leader of the Irish Sein Fein (IRA) party, Gerry Adams, made a vacationing visit to Cuba. Adams was warmly greeted by Fidel. Then, in a rambling speech, Adams declared that the IRA's struggle to evict the English from Ireland was the same struggle that had been waged successfully by Castro in overthrowing the Batista government. Those remarks were incomprehensible to the world's press reporters because they could not imagine what Gerry Adams and Fidel Castro had in common. What they may truly have in common is that both are Jesuit revolutionaries; fighting to overthrow their common "oppressor" i.e. Protestant Capitalism.

In the United States, the Papacy, through its Jesuits, openly supports the Marxist agenda. Its influence over Catholic politicians explains the knee-jerk liberalism of such prominent figures as Senator Edward Kennedy, Senator Daniel Moynihan, Senator John Kerry, Senator Dodd, Senator Leahy, Senator Hillary and ex-President Bill Clinton. Following the Papacy's lead, these men and woman intend to convert the United States into a godless dictatorship; *by legal means if possible but by revolutionary means; if necessary.* Their subversive political power is continuously being brought to bear upon the U. S. Government and its bureaus from that nest of Jesuitism; Georgetown University in Washingto, D. C.

CIVIL RIGHTS

The most important leader of the African-American Civil Rights Movement was M. L. King. Nominally, he was a Protestant Minister from a small church in Alabama but, mysteriously, he was educated (on a scholarship no doubt) at Boston University which is a Jesuit institution. There he imbibed the dogmas of the Jesuits and was trained to be their tool. It would appear that in his heart King was sincere in his beliefs about advancing the status of his people but, nevertheless, he was totally under Jesuit "control"

Roy Abernathy, the movement's second in command, recently told a story about the "March on Washington" that culminated in the famous "I Have A Dream" speech, by King given on the steps of the Lincoln Memorial.

Abernathy was scheduled to speak first to introduce King and he had prepared a written speech that included the usual demands for political and social equality. But, recently, he revealed that, some hours before the time for him to go on, his prepared speech was taken from him and rewritten by no less a personage than the Catholic Archbishop of Washington, D. C.! When the speech was returned to him, it now contained added paragraphs outlining veiled threats against the Congress if it did not pass the Civil Rights Bill; then pending before them. It implied that Congress *must* pass the bill—or else. "Or else," meant the threat of more riots. This revelation, coming from Mr. Abernathy's personal testimony, must be a true story. It proves that behind all those years of riots and marches that tore apart the nation under the banners of Civil Rights, there was a secret hand at work; the hand of the Jesuits. In truth, PAGAN organized and financed the whole Civil Rights Movement; every step of the way. Then, when their puppet leader was no longer needed, they pulled him off the stage by direct action.

THE CIA

The most important Jesuitical organization within the U. S. Government is the Central Intelligence Agency. For 60 years, this agency has waged a series of military and subversive wars against many nations around the world on behalf of the Papacy. Founded by a Catholic, Colonel (Wild Bill) Donovan, it has always been under Jesuit control and has been led by Catholics. (i.e. the Dulles brothers, Allan and John Foster, whose nephew now is a Cardinal.) Both its budget and its aims are secret; secret that is, from the American people who must pay for them. Around the world this secret army works, hand in hand, with the local Catholic priests in each country in carrying on its espionage activities against foreign governments. (see the book: "My Spy" in the bibliography.)

Worse yet, while waging secret wars against innocent nations, it has descended into the use of hideous methods of torture to extract information from prisoners captured by our military units. These tortures have often ended in the death of the victims. These atrocities conducted outside the law of nations and of mankind, have shocked the American public to its very core. Never in any manner has the Congress authorized such methods. Yet, the CIA, and its surrogates continue on. Involved in this practice, now, are the U. S. Army and the U. S. Marine Corps who, thereby, have forfeited any claim to honor. They have become an Army and a Corps of Dishonor. They are an Army and a Corps of disgrace. How can they let the CIA continue to do this in areas of their responsibility? The CIA/Jesuit interrogators are insane beasts who believe they are above the law as agents of the Jesuit's diabolical Inquisition.

THE HANSSEN AFFAIR

A further example is the case of convicted FBI master spy, Robert Hanssen. Employed by the FBI in Washington, D.C., Hanssen, it was alleged, supplied highly sensitive data to the Russian government, for money. He did this for over twenty years and yet, somehow, he went undetected. In his defense it was said of him that he *was*, at least, a good Catholic because he went to Morning Mass almost every day.

Even when his wife told their parish priest that she thought her husband was spying, the priest failed to report her suspicions to the U. S. Government. Perhaps, he took Hanssen into the confessional booth, gave him absolution and told him to "go and sin no more." So much for the loyalty of our American Catholic priests and of our American Catholic Church!

Or was there another reason for Hanssen going to Mass so faithfully? Perhaps the true story is that he was not supplying information to the *Russian Government* but was in reality delivering his top secret, stolen, information to the *Vatican Government,* through this Catholic priest, who then sent it on. by diplomatic channels, to its true destination—Rome.

What is the truth? We will never know because the truth about what the CIA and the FBI have done, and are doing, is totally hidden from the public behind a wall of secrecy; hidden by a government that asserts it has the right to lie to the American people. This use of secrecy will, eventually, bring on tyranny. We have become a Government of Lies.

*   *   *   *

During its first millennium, the Catholic priesthood lived as others lived; free to marry or not to marry as they wished. Only those who voluntarily chose the monastic life renounced marriage. But, as the secularism of the Early Renaissance began to affect the thinking in Europe, the Papacy, to more rigidly thought-control its priestly army, in 1139, instituted the rule of celibacy for all; priests nuns and monks alike. By the enforcement of this rule it succeeded in divorcing priests from their personal attachment to their wives and children. Thereafter, their legitimate family could only be their fellow priests. In the Papal Plan, celibacy is the cement that binds the system together.

"Celibacy" however means one thing to the public and quite another to the Pope and his priests. Understanding the undeniable demands of man's carnal nature, what the Vatican actually means by "celibacy" is that priests or nuns cannot legally marry. Otherwise all else is permitted.

On the occasion that a priest and a nun fall in love and decide they should marry, both are instantly dismissed from their orders. On the other hand, if a priest is discovered in an adulterous relationship with one of his female parishioners, he is protected by the Church and, even if found to be guilty, is not dismissed. He is merely transferred to another parish until the scandal in the old parish simmers down. In those cases, the Church is silent because the priest has really not violated his vows.

In recruiting new priests, it is common as the Catholics say they choose "the best and the brightest" at the tender age of 13 and then lock them up in the all-male atmosphere of a Junior Seminary. There, driven by the physical demands of puberty, but denied the outlet of female companionship, many, if not all, are led into the practice of homosexuality, pederasty and other perversions. Beyond doubt, these secret vices have been common for centuries but now, with advent of HIV/AIDS, they are being widely exposed.

A recent newspaper article revealed that, in the United States alone, hundreds if not thousands of Catholic priests have died of AIDS since that plague began in the mid-1980s. Obviously, they could not have acquired the HIV virus from the practice of celibacy. Yet, even with such evidence before them, the Church makes no response, or any attempt to change. In the face of these damning facts, its rules and dogmas remain immutable. Their lips are clamped shut. How many are living with AIDS is not known. How many more are still practicing homosexuals is an unanswered question. (Judy L. Thomas. The Kansas City Star January 30 2000)

*　　　　*　　　　*　　　　*

In a recently published book Donald B. Cozzens, Rector of St. Mary's Seminary in Cleveland, Ohio, deplores the failure of the modern Church to recruit and retain new priests. This has resulted in many parishes being severely understaffed. He goes on to admit that probably one-half of the priests in service are homosexual and that in the seminaries probably more than one-half. Yet this article is an amazing illustration of the Catholic mind-set in that this author does not recognize that the issue of homosexuality and the difficulty in recruiting new priests are interconnected; that the reason so many reject their call is because they reject the demands of celibacy and do not want to be a part of it.

The rule of priestly celibacy is at the very heart of the wickedness and dishonesty which the Catholic System represents. Were this single rule abolished it would, on the instant, begin a reformation within the Catholic Church

Cozzens Donald B. "The Changing Face of the Priesthood." Liturgical Press 2000.

*　　　　*　　　　*　　　　*

Recently, a learned historian and economist, after a lifetime of analysis of the world's political systems, says he has come to the conclusion that the essential prerequisite for any free society is the institution of legalized private property. A review of the social scene around the world confirms his conclusion. What he fails to state is that the right of individuals to own private property only came into being as a corollary of the Christian believer's individual salvation by Grace. Only under Protestant governments do individual property rights have any meaning.

In "developing" countries (the word "developing" being double-speak for the true word which is "collapsing" countries) the heart of their problems is that here there is no common agreement among the people with their government that titles to property have any meaning. Under European colonial rule, in Africa, many fairly efficient economies were built up. In Zimbabwe, (formerly Rhodesia), the economy was based upon specialized agriculture on farms largely owned and managed by white Europeans. When the British left, it was widely expected that at least Zimbabwe would remain a successful economy even with a native government. But what happened in Zimbabwe, as in most the other new nations in Africa, was the imposition of a reign of pyramidal Socialism that virtually abolished property rights. The government of Zimbabwe is now totally canceling the right to private property. The result is universal ruin. Without the right to own property, the people (White or Black) will not work, and when they do not work, they do not plant the fields. The result is famine in a fertile land.

Disastrous as these policies obviously are, they are continued on by Zimbabwe's President, Robert Mugabe. He and his government thugs continue to drive the white farmers off their land. Their farms are then given to black squatters. But since these do not know how to operate them, the farms fall into ruin. This is done in the name of social justice. But one must ask the question: justice for whom? While the white farmers are dispossessed, the natives are starving. Why is this? Why does Robert Mugabe carry on with this obviously failed policy?

An answer to this question was recently revealed; when, upon the death of Pope John Paul II, Mr. Mugabe was seen to be present among the notables at the Pope's funeral in Rome. A news article in the British press has revealed that Mugabe, like Fidel Castro, is a devout Catholic. Suddenly, his program to destroy Zimbabwe's, private-enterprise, economy makes sense. It explains why most of the world's governments continue to support him. De facto, Mugabe is the Catholic Archbishop of Zimbabwe! Very likely, he, like Fidel Castro in Cuba, is a Jesuit priest and is therefore untouchable. What he is doing to Zimbabwe is carrying out the Vatican's revolutionary social plan; which is to eliminate Free Enterprise, Private Property, and the Middle Class in Zimbabwe; and to establish the "new feudalism" in Central Africa. "The Week" Magazine; April 22 2005. p. 8."

    *   *   *   *

NEXT-AN AMERICAN GESTAPO?

News Item: March 21 2001 (Six months before the World Trade Center Atrocity), "Congressman Mac Thornberry, R., of Texas, introduced legislation (HR-1158 107th Congress 1st Session) to create a National Homeland Security Agency, a super-agency, to coordinate and to bring together several existing agencies for the purpose of combating potential terrorism." If it were not so frightening, the name chosen for this anti-terrorist agency would be laughable because it copies exactly the name of the NAZI Gestapo which in German was called the "Geheime Staatspolitzei" The Homeland State Police! Apparently PAGAN believes that if a name and a program worked once, there is no need to re-invent the wheel. If and when this super-agency becomes established who are the "homeland" terrorists it will combat? From its name it must be concluded it will be attacking "internal terrorists" who might be anyone whom they decide opposes their rule. In the mind of PAGAN (i. e. the Jesuits and the Vatican), an Office of The Inquisition is always a practical necessity. (www.asia.dailynews.yahoo.com) 3/28/2001.

*   *   *   *

September 11 2001

After 9/11 All true Americans mourned for the victims slaughtered in this appalling tragedy. But as students of realpolitik, we must seek out the true facts and discover, if we can, what really happened on that fatal day: September 11, 2001, and what might be done about it.

Was this tragedy a surprise to our government or did it *know.* Did it know these atrocities were being planned? One significant fact is that on August 1, 2001, a full six weeks before the World Trade Center event, all U.S. military installations were abruptly ordered into a high, state of alert. The bases were locked down under new rules of security. All persons entering the bases whether military or civilian thereafter had to have new passes and present new types of valid personal identification. After that date, everyone had to stop and be checked out each time they entered the posts. No longer were people and automobiles waved-in through the gates.

This suggests that the government *knew* the disasters were being planned and that it might even have been the force behind them. As unbelievable as this seems, in the past when deemed necessary, similar atrocities have been committed by governments against their own citizens under the orders from the highest authority, PAGAN. Is this one more example of the "The Red Terror?"

In 1915 no one could believe that the British government would arrange for the sinking of their own ocean liner, the Lusitania, for the sole purpose of bringing America into the Great War; but it did!

In 1933, no one in Germany could believe that Hitler would burn the Reichstag in order to create a political crisis demanding he be made dictator; but he did!

In 1941, Americans never imagined that President Roosevelt, who knew in advance, the date of the coming attack, would permit the Japanese Navy to bomb Pearl Harbor and sink our Navy's ships while they were helplessly tied up at their docks deliberately prepared for sinking; just to create an atrocity big enough to force the American people into supporting our entry into World War II; but he did!

And in the year 2001 AD, no one in America could imagine that any U.S. government official would take part in the destruction of the World Trade Center; but did they? Certainly, no ordinary American would participate in such a plot. But what are the alternative possibilities. Did any of the poverty-stricken Muslim "enemies" of America have the technical ability to orchestrate such a complex operation and carry it out without it being discovered beforehand? It seems extremely unlikely. Our intelligence agencies do not fail. They are nearly omnipotent. They knew what was going to happen and warned their own friends in time. (Viz. Condalezza Rice et al.) What organizations would commit such an act as a necessary precedent to achieving its goals?

PAGAN i.e. the Imperial Vatican State; would!

The Society of Jesus, the Jesuits; would!

Our limp-wristed, Jesuit-trained, Foreign Service officers, if ordered to, would!

The Illuminated Freemasons in the United States Congress; would permit it and say nothing!

The Zionists of both America and Israel; would!

Those organizations all agree upon one key principle: "that the end justifies the means!"

If this thought could be true, their motive might have been to persuade the stunned American people into surrendering the last shreds of their personal liberties by accepting the creation of an American Gestapo. Yes, that same **"National Homeland Security Agency"** that was "trial ballooned" by Congressman Thornberry six months before. Eerily, this is the same name in English as the one used by the Nazi Gestapo in Germany. (Das Geheimlich Staatspolitizi= "The Homeland State Police")

Three days following this epic tragedy, on (D+3) September 14, 2001, several politicians including the President and Senator Hillary Clinton mentioned the word 'Homeland" in their TV interviews. The next day (D+4) September 15, 2001, the Associated Press issued an article citing the need for "Homeland

Defense." Also, on that same day, (D+4), Saturday, September 15th, a televised conference was held in Washington, DC. It was a panel of experts who proposed the establishment of (guess what?) yes a **"National Homeland Security Agency."** They stated that their Study Commission had been working on this plan for the last two years! The conference was moderated by the chairman of the Council on Foreign Relations (CFR) and was attended by most of the members of Congress. Included on the study panel were two disgraced former politicians, ex-Senator Gary Hart and ex-House Speaker Newton Gingrich. (By their fruits ye shall know them)

WHO KNEW?

On September 17th, 2001, a reliable business news agency reported that beginning on September 6th (D minus 5) thousands of "put option" contracts were bought in the New York stock market on the stock of United and American Airlines. Put options are bets that the underlying stocks will go down in the near future. The number of put option contracts sold per day was 10 times the normal level of activity. This is conclusive proof that **someone knew** that a major assault was coming and that it would negatively impact, exactly, those two airlines. Does this means that someone knew there was an official plan to destroy the World Trade Center by using hijacked airliners and that information about the plot had been leaked to someone greedy enough to try to profit by it? Was this an Islamic Israeli or American investor? To date no one has been identified as the buyers of those options but certainly there are records. What *is* the answer? (Bloomberg Business News 9/17/01)

Subsequently there was no follow up on this information and there seems to be no official desire to discover the names of these profiteers. The story has, now, been dropped down the "memory hole" and the truth will, likely, never be known.

CNN TELEVISION NEWS

During the afternoon of September 11, 2001, while scenes of the disaster shown on television had Americans in shock, CNN broadcast a visual report, ostensibly being taken, "live," somewhere in the Middle East. This was a film clip showing Arabic crowds dancing in the streets; celebrating the fact that America had finally been punished. But it turns out that this CNN report was a vicious lie. Alert viewers recalled that this film clip was actually 10 years old! What it really depicted was Palestinian citizens celebrating the U.S. military's liberation of Kuwait from Saddam Hussein during the Gulf War in 1991! The broadcast of this deceptive film clip was a deliberate lie and means that the CNN news editors *knew*, in advance, that the 9/11 atrocity was coming and they were ordered to res-urrected this film clip, in advance, to have it ready to be shown immediately after the event. This lying film clip was a veritable "propaganda bomb" dropped upon the American public. It was shown, specifically, to begin the creation of the offi-cial "Cover Story;" the Cover Story that this atrocity had to have been the work of "Muslim Terrorists." From that day to this, the United States Government's official position has remained the same; that the perpetrators of these crimes of mass-murder had to have been "Muslim Terrorists." In fact, our whole foreign policy from that day to this has been based, precisely, upon that premise and upon that, faked, CNN, film clip.

∗   ∗   ∗

9-11/AN EYE-WITNESS REPORT

On September 11, 2001, just after the collapse of the World Trade Center Towers and with clouds of dust still rising in background, a local NY television station's news cameraman was down in the streets interviewing people who might have witnessed the event. One intelligent looking young man was asked: "Did you see what happened?" and this man replied: "Yes, he had seen the plane that hit the second tower." But, he said, "it was not an airliner; it was smaller (smaller than an airliner) and it was all-gray colored; with no markings." At this, the startled reporter said "NO, NO, it was an airliner, it was an airliner!" The announcer, apparently, understood that this information was too hot for him to handle.

Could the planes that were used to hit the towers actually have been radio-controlled, smaller, drone airplanes; directed to fly into the buildings by remote control? Consider the physical problems that would have been faced by the inex-perienced "Muslim" pilots in flying, huge, Boeing 767 airliners, accurately, down into targets as relatively small as a single office tower in Manhattan? Then, imag-ine another set of inexperienced "Muslim" pilots being able to do it again? That is nearly impossible. Experienced airline pilots would hardly have had the ability to do this. They land their airliners only on huge airports with the aid of sophisti-cated ground control assistance. However, remotely-controlled drone airplanes would have been be far more accurate and certain of success

Nevertheless, the crashing of the airplanes into the two buildings was merely a diversion; a diversion to mask the true cause of the buildings' eventual collapses; their demolition by means of previously installed internal explosive charges.

WERE THEY CONTROLLED DEMOLITION?

One thing is certain; the crashing of airplanes into the Twin Towers did not cause them to fall because, after the impacts, both, Towers remained standing. But then, after about 30 minutes, both began to collapse—straight down. Structural engineers say that if the collapse of the towers had been caused by aircraft collisions or from the resulting fires they would not have fallen straight down as they did. If the fires had caused the collapses, the tops of the buildings would have toppled over sideways; but they did not do this.

TV pictures of the buildings falling clearly show that their upper portions moved straight down upon the lower floors. The line of failure was equal all around the buildings at about the fiftieth floors; far below the 84th floors that suffered from the collisions. That straight down movement could only have been caused by a controlled, explosive, demolition with all of the bearing piers on each floor being blown out simultaneously. In the videos, one can see the puffs of white concrete dust being blown out on all sides, all at once, from the successive of the floors as their piers were blown out. Then, with no support, the upper portions of the buildings began to fall, inexorably downward, under the force of gravity until they became giant pistons; crushing the floors below; one by one. This is what the TV pictures, actually, show us.

At the same time that Towers One and Two were falling, another building, Building Six, mysteriously imploded as well; and for no apparent reason. It had not been struck by an airplane.

But how could all these buildings have been prepared for demolition? It could only have been accomplished by assuming that the buildings' operators would permit it. The leaseholder and operator of the WTC complex was, and still is Lawrence Silverstein, a Jew, who, only 6 weeks before, had bought the World Trade Center under a lease-purchase agreement from the Port of New York Authority. It is said that he was not finding the project profitable, but even so, he insured the buildings for the astounding amount of $3.5 billion dollars. Then, probably in response to a secret governmental agency's demand, (the CIA, NSA, NRO or whomever) he turned over several empty floors in each building to certain U. S. or Israeli "demolition experts" so they could set their explosive charges without being noticed. As the buildings' operator, he could not care less; because he would get the insurance. Financially, he could only gain by their destruction. Nevertheless, still being loyal to his fellow Jews, he let it be known through their private communication network that none of them should come to work on D-day. It is on record, that none of the 3000 victims were Jewish; although, normally, thousands of Jews worked in the buildings. Of course, the deaths of 3000 "goyim" would be of no concern.

On September 11th, as the buildings were falling the "demolition experts;" supposedly a group of five Israeli "furniture movers," were found to be on the roof of a moving van some miles away in Liberty Heights, New Jersey, video taping the events for their masters; all the while dancing and cheering as the buildings were coming down. What could they have been cheering about? What they might well have been cheering about was the success of their demolition work. Is it not possible that these so-called "furniture movers" were, in actuality, members of an Israeli Mossad demolition team; sent here to do a job no American would perform? Why not? Travel between Israel and the U. S. is nearly unrestricted. Significantly the manager of the alleged "furniture moving company" whom these "movers" nominally worked for, fled to Israel the very next day when his "moving men" were arrested. Later these "dancing Israelis" were held incommunicado in a Federal prison for several months but they were never charged with any crime. Later, after the public had forgotten about them, they were quietly released and flown back to Israel.

HOW WAS IT DONE?

Technically how could the supposed Islamic hijackers have had the skill to fly two large airliners into each of the Trade Center Towers and into the Pentagon building in Washington D. C. as well? For years, experts have discussed the pros and cons of this question. One important fact that has become fully apparent is that whatever hit the Pentagon it was not a Boeing 757 because the damage was not significant enough to have been caused by such a large airplane. The damage that did occur was only consistent with the impact of a small single engine plane or a single guided missile. Significantly, and of conclusive proof, is that only one small jet engine was ever found within the wreckage.

But what about the accuracy of the planes' guidance systems; assuming they were not flown by human pilots? Recently we have discovered the likely explanation. It is from privately gained information that reveals the targeting method used by the U. S. Special Forces as they infiltrated into Baghdad, before, the American and British armies moved forward in March, 2003.

THE ANSWER

A technical question concerning both the Iraqi Wars, I and II, is how could our, Navy-launched, cruise missiles have been guided so accurately that they were able to hit specific buildings in Baghdad after flying for hundreds of miles? This degree of accuracy seems incredible. We have been told these missiles were guided to their targets by our Global Positioning Satellites. But, even with their aid, how could they have picked out single specific buildings? The answer is that our undercover agents, dressed as Iraqis, infiltrated into Baghdad; carrying with them what looked to be rolled up advertising posters. To the natives, these would appear to be just that, ordinary posters; but these posters were not for advertising; they were targeting panels made to be pasted up onto designated buildings marking them for destruction. They carried Arabic ads, but were made with a special reflective material or perhaps they contained the new Radio Frequency Identity Chips (RFIDs) that can be glued onto paper and operate as radio reflectors or re-transmitters. These chips are the similar to the RFID chips now beginning to be used to identify merchandise in our super markets, or that pet owners and parents are now injecting into their pets or children to identify them when they are lost. Once, one of these posters was pasted upon a building, that building was doomed. As a cruise missile approached, the posters transmitted back their own unique identifying signals; thus providing each missile with the location of its designated target. Upon recognizing it, the missile would then home in and destroy the building at a blow. One target—one missile.

Might the use of these RFID chips not explain, also, how two drone airplanes could have been have been guided into striking not just one, but both, of the World Trade Center Towers? And might they not also explain how another plane, or missile, could have been guided into the Pentagon as well? All that would have been necessary would be to have to have pasted one of these targeting panels onto the exterior of each of these three buildings. With those in place, the drones or missiles would, upon recognizing the panels, fly into their targets without fail. In fact, the TV films of the crash of the second plane into Tower I, shows that the plane almost missed the building. In the final moments it banked at an almost impossible angle, with its wings 90 degrees to the horizontal; desperately trying to hit its target. In the event, it just barely nicked a corner of the structure.

LATER EVIDENCE

In the June & July, 2006, invasion of Lebanon by the Israelis, Lebanese authorities reported that they had discovered members of an Israeli secret spy ring operating in Beirut that were going about pasting up "signs" on buildings. They did not explain if they knew why these signs were being posted but, without doubt, these were providing guidance to the "smart-bombs and guided missiles that were being fired into Beirut from Israeli, F-16, fighter planes; those very planes and missiles supplied to them by the United States government.

CONTROLLED DEMOLITION

Nevertheless, the crashing of the aircraft or missiles into these three buildings was only to create a diversion; a diversion to mask the true cause of their destruction. The real cause of their destruction was their controlled demolition by previously placed explosive charges. These massive atrocities were created entirely to establish the government's pre-arranged "Cover Story" that was to incriminate alleged "Muslim Terrorists."

It has been asserted by one engineer that the demolition charges in the World Trade Center had actually been installed as the buildings were being erected, and that they were there from the beginning. Therefore, to explode them, it was only necessary to attach the necessary wires and close the electrical switches at the proper time; thus creating a total collapse of the structures. It is said this was done because, in looking to the future, the buildings' architects and engineers saw that, eventually, the buildings would have to be torn down and that the cost of disassembling these 100 story buildings, piece-by-piece, while protecting other nearby buildings, would then exceed their original cost. In today's dollars it would have cost a billion dollars. So, to make the eventual demolitions feasible, they determined that the only way to demolish the buildings after they had reached the end of their useful life, would be to bring them straight down by a controlled demolition.

The truth is, that by the year 2001, the buildings were already economic white elephants. They had been built by the Port of New York Authority, solely as prestige symbols, without regard to their real economic value. The buildings were not well built and now it was found they were, also, full of asbestos insulation.

Therefore, when it was time to stage a monstrous atrocity for political purposes, what better buildings were there to destroy than these buildings after insuring them to the maximum, to then bring them down and convert these economic liabilities into billions of dollars of cash!

FINAL PROOF:—THE STORY OF BUILDING SEVEN

If any is needed, final proof that the 9/11 atrocities were not the work of "Muslim Terrorists" is the true account of how a fourth World Trade Center building was destroyed. The media does not mention it now but this was the demolition of "Building Seven;" better known as the U. S. Customs House Building. This was a third massive building in the WTC complex. It was forty-seven stories tall and covered about one city block. It was located away from the Twin Towers. Like the other buildings, nearby, Building Seven was evacuated during the morning of 9/11 but remained unharmed. As late as four o'clock PM, on September 11

2001, it was still intact. But then, without warning, at about 4:30 PM, it suddenly collapsed into rubble for no apparent reason.

Later, in a statement, made in 2003, on a PBS television program entitled "America Rebuilds," Mr. Larry Silverstein, the Jewish operator of the whole complex, said that, in conjunction with New York's Mayor Louis Giuliani, a Catholic, concerning Building Seven; that they, jointly, decided to "pull-it!" When they gave that order, this perfectly sound building had its supporting piers blasted out at its base and it collapsed into a heap of rubble; in the same manner as Towers One, Two and Six.

This recorded public statement provides us with "proof certain" that Building Seven had also been sabotaged with explosive charges just as had been the three other buildings; Towers One Two and Six, IN ADVANCE, and that these two mass-murderers knew it! Concerning Building Seven, what probably happened during the morning of 9/11, was that there had been a critical failure. The third airplane (?) had failed to arrive; and/or the charges in Building Seven, programmed to go off along with those in Towers One and Two and Six, had somehow misfired. Whatever, something awful (from their point of view) had gone very, very, wrong. Now, at 4:00 PM, the deadly charges were still set in place inside the building. What were Lawrence Silverstein and Mayor Giuliani to do? After, one can only imagine were intense consultations with their masters in Washington, D. C.; or with FEMA Headquarters in Manhattan, they were finally ordered to make a "command decision" to act. Rather than have anyone enter the building and find it was mined with explosives, Larry Silverstein, the person responsible for getting the "job" done, and Mayor Rudolph Giuliani, his accomplice, decided they had to "pull-it." That is, to manually close the switches to set off the charges. They were forced to do this because it was absolutely essential to conceal from the American public the awful truth that all of the 9/11 atrocities, both in New York and in Washington, D. C., were not the work of "Muslim Terrorists," at all, but that they were a deadly hoax played upon the American people; planned and executed by their own United States Government and by PAGAN. In the process, they murdered some 3500 American citizens. It was all designed to persuade the American people to accepting slavery under the Homeland Security Agency and its *Patriot?* Act.

The fact that Building Seven was wired with explosive charges, placed there far in advance of September 11, 2001, PROVES ABSOLUTELY, that the World Trade Center Atrocity was deliberately planned and executed by someone or some agency. And it PROVES, ABSOLUTELY, that the official Cover Story is a lie.

By understanding the story of Building Seven, another thought arises. It is that the great loss of life that occurred during the destruction of Towers One and

Two might have been what is euphemistically called "collateral damage;" loss of life not contemplated in the original plan.

What may have been planned was to wait until the buildings had been totally evacuated and—then—to have exploded the charges to bring down the buildings. However, unforeseen circumstances forced a change in plan because, while the tenants and workers were descending the stairways to the street level, the New York Police and Fire Departments were invading the buildings. The firemen were climbing *up* the stairways to get to the fires and to extinguish them. In their last radio communications, before the city wide Police/Fire emergency radio system mysteriously went dead, they said they had reached the 35th floor and were reporting sounds of explosions above them. Therefore, they were only 15 floors below the 50th floors: the vacant utilities service floors where explosives were probably placed.

The controllers in charge of the demolitions well knew that if the firemen reached the 50th floors in either building and reported the existence of unexploded charges, the whole purpose of the operation would have been blown. The "Muslim Terrorist" scenario would have evaporated and future political events would have had to take a drastically different turn. If the explosives charges had been discovered and reported, the whole Muslim conspiracy story would have been blown including any justification for our pre-planned offensives against Osama bin Laden, Al Qaeda, Afghanistan, Saddam Hussein, Iraq; or the need for a *Patriot Act*—everything, everything, everything—blown!

Faced with this ghastly possibility, from their point of view, PAGAN made the first of its two "command decisions" on that fateful day. It ordered the switches pulled on the explosive charges, all at once, in both buildings; *before* the firefighters could discover and report their existence. Truly, the New York City police and firefighters, in their devotion to duty, were more than just heroes, they were martyrs; martyred by PAGAN

Consider these proven facts the next time you go through an airport security strip-search or when you find that your phones and e-mails are being tapped; or when, in extremis, you should stand before one of their newly invented "Tribunals" whose procedures are faithfully copied from those of the Spanish Inquisition.

THE ANTI-TERRORIST PROGRAM

On September 20 2001 (D+9) while the American public was still bewildered and hoping for answers to the "terrorist threat," President George W. Bush gave a major address to a joint session of Congress. In it he stated he had established a new agency to fight "terrorism" called, not surprisingly, **The National Homeland Security Agency!** (Quid est Demonstrandum)

If the reader has not viewed it yet, they should rent a video of the movie "Wag the Dog." It is very instructive. The book "1984" is even more profound.

*   *   *   *

CARNIVORE

On (D+3) September 14 2001 the US Senate passed a new act "**The Combating Terrorism Act.**" Under this bill, a bill that many Senators did not even bother to read, Federal Prosecutors will now be enabled to authorize surveillance of any form of communication, inside the United States *without* a judicial order. The Federal government has already built two powerful computerized surveillance systems named "Carnivore" and "Echelon." These, it is said, can simultaneously listen to and record thousands of telephone conversations, E-mails, Internet chat-room conversations, and can even record what sites one visits on the Internet. Quite possibly, Carnivore or some other system might be able to enter into and read the files on the hard disk of your personal computer. Whether they can really do these things; the fear that they might, will act to stifle free speech, free thought and free communication. The bill was brought to the floor of the Senate and forced to a vote with only 30 minutes of debate! Most of the Senators had not even read it. A Mormon, Senator Orrin Hatch, of Utah, was its sponsor.

On October 12, 2002, the House of Representatives passed the Senate/House Compromise Bill HR: 2995, 107th Congress, First Session. It was renamed "The USA Patriot Act." The bill was passed with a majority of Republicans voting for it while many liberal Democrats vehemently opposed it. These advocates of "civil rights" now, only too late, now realized that the passage of this bill meant an end to "civil rights."

The final version of the USA Patriot Act was passed by the Senate on October 26, 2002, and was numbered Public Law 107-56. After signing it, President Bush appointed, as the Homeland Security Agency's first director, Pennsylvania's governor, Thomas Ridge. No doubt Governor Ridge is an honorable man but it would appear that he is, also, a Roman Catholic.

*   *   *   *

PERMANENT WAR EQUALS PERMANENT PEACE!

Following an end of the synthetic Cold War the United States government was psychologically without a serious external military threat. We were triumphant.

Because of this the nation was without a sense of "purpose." This lack has now been remedied. "International Terrorism" is to be the new enemy against whom we will, perpetually, make war. Conveniently, it is a war that needs never be won. "Terrorism" will maintain the country in a permanent state of fear. This fear will strengthen the role of the government by forcing the common people to cling to it ever more tightly as their only apparent defender. The "War against Terror" will justify the invasion of any uncooperative state, anywhere in the world, in the name of "fighting terrorism.' Those countries that do not surrender will be pulverized by bombings until they see the light and agree to be incorporated into PAGAN'S One World Dis-Order.

The motivation for the U.S. invasion of Afghanistan may well be that it was not really waged against "Terrorism" but that it was waged to eliminate the Taliban government who, throughout 2001, refused to agree to the building of an oil pipeline through their country. The proposed pipeline was to bring Russian oil from the fields of Central Asia down to the Indian Ocean. This would of course further enrich the British-American oil cartel. It is reported that early in 2001 British, U. S. and Russian oil negotiators offered the Taliban two choices: "Either we will pave your country with a carpet of gold or we will pave it with a carpet of bombs." The Taliban leaders chose the latter course.

The next attack, the unprovoked assault upon the nation of Iraq in 2003, was a further step in the conquest of more oil reserves and of containing or converting the Moslems; a fixed Vatican goal. The United States and Britain intend to occupy these countries with permanent military bases to maintain their puppet regimes. The American and British people give little support for these wars but the United States President George Bush II went ahead and commenced wars against these sovereign states, anyway, in direct violation of the U. S. Constitution. Tony Blair, the Prime Minister of Great Britain, followed suit. But why would Tony Blair go against his peoples' and his own Labour Party's wishes? He did it because he is a crypto-Catholic and thus he obeys the Vatican's will. This fact was confirmed in 2001 when Blair accepted communion from the hands of Pope, John Paul II, and by his attending the Pope's funeral in 2005. There is no doubt but that Tony Blair is the Vatican's man. Here is a quote from The London Daily Telegraph; reprinted in "The Week" magazine April 22, 2005. p.18

> "It wasn't just Blair's star-struck admiration for the pope that prompted his shocking breach of protocol said Damian Thompson in the London Daily Telegraph. "Blair is secretly a Catholic. He may still claim to be an Anglican but his true religious devotion is to the Vatican. Most Sundays he attends mass with his Catholic wife and kids. At the pope's funeral, while Prince Charles merely leaned forward

awkwardly in the time-honored manner of Anglicans who cannot quite bring themselves to kneel in a Catholic church, Blair did the whole genuflection-obeisance thing. His grief was surely real. After all, though few people know it, Tony Blair took Communion from Pope John Paul II, himself, at the Vatican in February 2001. He was the first British prime minister *ever* to receive the sacrament from the hands of a pontiff. So much for our legal establishment as a Protestant nation."

In truth what we see in Iraq and Afghanistan are not the United States' or the British armies. No. In reality, what we see is that they are Papal Armies; poised on the Southern border of Russia. Their next thrust may well be Northward in an effort to destroy the political and spiritual power the Eastern Orthodox Church and to replace it with Roman Catholicism. This is after all the ultimate goal of those madmen in the Vatican; the goal of fulfilling the prophecy of the Third Secret of Fatima!

INTERNAL CATHOLIC DISSENT

In present-day Catholicism there is a division between some who call themselves "progressives" and others called "traditionalists" as to how the Church should be organized. The progressives or "pluralists" argue that decisions on doctrine and practice should be decided by the laity and their bishops in their local areas; perhaps in consultation with the Pope; but considering also the will of the membership. In short they want the Church to be democratized and made more representative; accommodating to present day realities. They propose a reconsideration of the Church's positions on divorce contraception and an end to the rule of priestly celibacy. The Progressives are in rebellion, yes, but they only desire change from within.

The traditionalists headed by the Papal authorities are of course opposed to almost any change. While they pretend to hear complaints from below they remain adamantly opposed to any material alteration of their pyramidal system that has worked so well, from their point of view, for hundreds of years.

Recently two books have been written by Catholic authors addressing these problems. One is English and one is American. John Cornwell "Breaking Faith; and Garry Wills "Papal Sin, Structures of Deceit." Both advocate the progressive position. While claiming to be sincere Catholics they document the many grave problems in the present-day Church. They oppose the dictatorial nature of "rule from the top" exercised by an "infallible" Pope. They question the refusal of the Papacy to reconsider the role of women in the church; they question the rule of priestly celibacy that has caused so many to go astray; they question the rules governing Catholic marriage and divorce and they favor permitting the laity and the local priesthood to participate in local decision making.

Unfortunately, one thing these very sincere Catholic authors have not yet faced is that the Church *cannot* change because it is at heart not a religious organization but an authoritarian *political* nation state. It is the "Second" Roman Empire. It is not a church but a political conspiracy; it is a totalitarian dictatorship organized upon military lines with its priests as its officers and the laity as its soldiers. It *cannot* change because if it gave way on any of these points it would lose its centralized control and self-destruct. The Papacy's dilemma is that if they make any changes, now, it would be an admission that the Church has been "wrong" in the past. But since the Church they maintain has always been "right" and "infallible" change now is logically impossible.

It cannot permit its priests to marry because their obedience would be compromised by the demands of their wives and children.

It *cannot* ordain women because women will not blindly accept the discipline required.

It *cannot* approve of contraception because that would mean fewer soldiers in the ranks.

It *cannot* permit decision making by the laity because this would split their Second Roman Empire into factions.

What John Cornwell and Garry Wills do not yet see is that they are walking down the same path once trod by another Catholic reformer, six hundred years ago, Martin Luther. When, as did Luther, they come to the end of that path they must conclude, as did Luther, that the Roman Catholic Church *cannot* be reformed from within because, being a political dictatorship, it cannot change. When they come to agree with Luther, they will come out of that organization; shedding its wickedness and become true (protesting) Christians.

*   *   *   *

PRIESTLY SCANDALS

Beginning in January 2002 a new development occurred in the public news arena. Perhaps even in response to this "Christian Overview of History" that reveals the political character of the inner "Church." Suddenly, PAGAN initiated a public self-expose' of the sexual failings of its deviant priests, bishops, cardinals and archbishops; both in America and elsewhere. Now, rather than being buried in the back pages of newspapers or suppressed altogether these charges were given public prominence. Everyone including Catholic newsmen and Catholic intellectuals joined in the witch-hunt to root out homosexual and pedophile priests. This appears quite strange because those practices have been common for centuries as a by-product of the sham requirement of priestly celibacy. Heretofore, while those crimes commonly occurred they were always concealed and kept secret. Why did the Church suddenly confess?

The reason may be to use the expose' as a smoke-screen to mask the political takeover of the United States Government by the Vatican's rigidly organized militant arm; the Society of Jesus. The exposure of the, of course guilty, (but lowly) priests is designed to present the appearance of a church in disarray; a church that certainly could not possibly pose a political threat to the independence of the United States.

The Jesuits expect these scandals will divert attention from them. However they themselves are not disorganized. They are preparing the final blow. After staging the recent 9/11 atrocities at the World Trade Center and at the Pentagon they believe they are poised to set up the One World Police State. In their plan of conquest they are abetted by their puppets in the Congress in the Administration and in the Supreme Court; each of whom thinks that he or she is only following the dictates of their own secret society's noble aims. Little do they realize that those societies were only set up to carry out the Master Plan of a higher conspiracy: World Rule by the Second Roman Empire. Only when it is too late will they discover they have been fatally deceived. Then, will there not be a "great wailing and gnashing of teeth" as they are each led off in their chains toward their respective Gulags.

*   *   *   *

THE GRAND DESIGN!

For the present, the Papacy, as the Second Empire of Rome, tolerates complaints from their dissidents as they complete their larger purpose which is to win their "War of the Counter-Reformation" against Protestant-Capitalism on the one hand and its war to block any further advance by Islam on the other. From their point of view they are fighting on two fronts. The borders of an aroused Islam are only a few hundred miles away from Rome to the East; while the latent power of the Protestant-Capitalist states lies just to their West in Northern Europe and North America.

After achieving total control over the governments of the United States and the European Union they plan to eliminate representative governments and to gradually reestablish, by force, their feudal religious political-socialist regime. Afterward, any dissidents from both within and without the Roman Catholic community will be summarily dealt with and Catholics will learn that the time of accommodation is over. Yes the new governments will be established as "democracies" but these will be democracies in which PAGAN controls the outcome of every election.

The definition of a "hate-crime perpetrator" or "terrorist" will soon become any person who opposes the dictates of the government. As the definition of a "hate-crime" is re-written tighter and tighter it will gradually bring down upon us all the final dictatorship of PAGAN.

This need not happen. But it is time for Christians who wish to retain their freedom to act. They have it in their power; but the hour is late. PAGAN believes it is gliding downward on an easy slope toward final victory. Let us make every effort to disappoint them. The way to block a conspiracy is to expose it and to conquer it by individually preaching the saving Gospel of Jesus Christ.!

January 29th 2002: President Bush in his annual State of the Union address to Congress proposed adding new powers to the Homeland Security Agency. These called for the formation of a citizen's "USA Freedom Corps" to act as the "eyes and ears" of the Nation. The USA Freedom Corps is to be brought under the direction of FEMA the Federal Emergency Management Agency. The USA Freedom Corps will then absorb and nationalize the already successful "Neighborhood Watch Programs." At ground level it will operate as "TIPS" ("The Terrorist Information and Prevention System.") This "TIPS" program will empower citizens to report to the government any suspicious activities they see, or think they see, in their local neighborhoods. Thus "TIPS" will become a universal espionage system encouraging American citizens to spy on and to report on one another. If fully implemented, this program will bind the American people in chains under a system of mutual fear and distrust. These Orwellian proposals indicated that we may have now replaced our Caligula with his natural successor Caesar the "Man of Steel;" "Big Brother."

$*$ $*$ $*$ $*$

PRIESTS ONLY UNDER CANON LAW

November 23 2002. In response to attempts by the American Catholic Bishops to establish new policies to deal with the sexual abuse scandals in the Church the Vatican and the Pope today declared that only *they* will be the judges of their priests in these cases and that all accusations will henceforth be tried by special Church Tribunals. These Tribunals shall try the cases under "pontifical secrecy!" The Pope thereby asserts that these deviant priests are not subject to the laws of the United States but rather that they are citizens of the Vatican State. Therefore their criminal acts will be kept secret from the civil authorities. This equals treason against the United States.

THE NEW INQUISITION

On May 18 2003 an article appeared in the Denver Post, and probably nation-wide as well, authored by two attorneys who outlined a new system of criminal courts called "Tribunals" or "Military Courts." These "tribunals" are to be used to try and to condemn "terrorists." Presumably the word "terrorist" may be defined as either foreign or domestic. What the article describes is appalling. It proposes a new draconian system of justice with none of the protections afforded to the accused under our traditional criminal laws. In the United States under our British-American system of Common Law any person accused of a crime has all of the following rights:

> The right to have his own attorney to represent him.
> The right to a writ of habeas corpus.
> The Fifth Amendment right against self-incrimination
> The right to a speedy trial.
> The right to be presumed innocent until proven guilty.
> The right to face his accusers.
> The right to a public trial.
> The right to a trial by jury.

Under the proposed new "Kafkaesque" tribunal system those protections are to be eliminated.

> Trials are to be conducted and judged by a panel of five military officers.
> The trials will be held in secret.
> The accused will have no right to know who has accused them.
> The accused will be guilty until proven innocent.
> The prosecuting attorney will be a military officer appointed by the court.
> The defense attorney will be another military officer appointed by the court.
> The defendant may hire his own attorney but that attorney will still be under the control of the military defense attorney.
> Conversation between an accused and his attorney will be recorded by the court. There will be no attorney/client privilege.

There will be no trial by jury.

The trial judges will determine the defendant's guilt or innocence and will prescribe the appropriate punishment.

There will be no right of appeal to any higher court. The condemned may only appeal to the Secretary of Defense or to the President whose decision will be final.

In presenting this program the authors build the case that the public must accept this new legal system because only severe methods can prove effective against international terrorism. Left unsaid is the thought that this same process may be applied to others. It may come to pass that anyone who questions governmental authority may be accused of supporting "terrorism" and given the same sort of merciless trial. Doubt it not; this may include both you and me.

Adding emphasis to the ideas floated in this article, the method of punishment to be applied to those found guilty is eerily suggested by a quarter page picture included in the article. It is an illustration of a guillotine machine. The inclusion of this picture could not have been accidental. There are rumors that such machines have already been manufactured and are being warehoused, right now, in military storage. These guillotines might be the government's new method for dealing with the need for executions en mass. And further, they should make a suitable impression upon the public's consciousness. As a method of execution the guillotine is efficient; it is swift; and causes little harm to the rest of the victim's body. The bodies might even be used a source for organ harvesting.

As a system of law this program is not new. It is the legal system long used by the Spanish Inquisition! Once one fell under the power of that hideous traveling court, anyone even suspected of failing to believe in the papal doctrines could be denounced by his own parish priest and then be dragged, in chains, into a trial called an "auto da fey." (a self-confession of faith). There, if it was decided by his inquisitors that he could not give satisfactory answers, he was judged to be a "heretic." When that dread sentence was pronounced, he was doomed. He was then turned over to the civil authorities and put to death. Our new inquisitors may view their proposed use of the guillotine as a form of mercy. During the first Inquisition, a period that lasted from the years 1100 AD until 1814 AD, the usual method was less humane. In those days the victim was tied to a wooden stake and burnt alive!

*   *   *   *

NEW CIA TORTURE PRISONS

Today it was revealed that the CIA has newly established secret prisons in foreign countries where they are practicing torture upon their accused victims that are illegal to use within the United States. The location of these 'black prisons' is secret—secret, that is, from the view of American citizens. Since this taxpayer funding of torture is now been made public, the President, the Congress and the Judicial Branch cannot claim ignorance of this return to barbarism. But they will do nothing because all are complicit; guilty of practicing torture by association. Instead, they are attempting to find and to prosecute whoever 'leaked' this information to the press. (Source: The Washington Post. 11/02/2005)

*　　　　*　　　　*　　　　*

SUPREME COURT APPOINTMENTS

In 2005 a key vacancy needed to be filled on the Supreme Court. After considerable debate Appellate Court Judge John Roberts was finally confirmed by the Senate to be Chief Justice. This occurred not simply because he was qualified by experience but because he was qualified, also, as being a Catholic; needed to replace retiring Justice Sandra Day O'Conner who was also Catholic. The unwritten rule in the selection process of choosing successive Supreme Court Justices is that they must be replaced; like for like; Jewish for Jewish; Black for Black and Catholic for Catholic. With this appointment the religious makeup of the Court is now five who are Catholics ensuring them a safe majority on critical cases.

To illustrate the mockery of the idea of any real separation between government and religion, on October 2 2005 the day after Justice Roberts was confirmed, he and President Bush, together with Cardinal Theodore McCarrick, Archbishop of Washington, D. C., attended what is called a "Red Mass;" at St. Mathews Cathedral. This is the traditional 'Red Mass' that is celebrated annually to honor of the opening of the U. S. Supreme Court's fall term. This indicates that the Judiciary as well as the other two branches is firmly under Jesuit control. (Associated Press 10/02/05))

*　　　　*　　　　*　　　　*

"PERMANENT WAR EQUALS PERMANENT PEACE"

This deeply wise quotation from George Orwell's book, "1984," accurately describes a necessary imperative faced by tyrannous governments. To maintain a state of psychological emergency over the minds of their subjects they must create a constant stream of threats or the actual reality of war being waged against them. Under a state of war, governments can then justify all types of illegal acts against their own people that otherwise would not be tolerated.

But to have a permanent war there is a problem. Suppose that, from weakness, the intended enemy surrenders and will not fight? This is a major problem. In war as in politics, or as in sports, there needs to be two sides.

In the 'wars' both in Afghanistan and in Iraq, those nation's armies were quickly defeated; so what then? From the point of view of our military perhaps it has been found necessary to organize "B-Teams" with whom to continue the war; B-Teams that carry on a so-called guerilla war against our own A-Team troops and against the civilians we pretend to be helping. One must ask; where are the bombs and weapons coming from? The Afghani and Iraqi natives have no ability to make these weapons. Neither do other Arab nations in the Middle East have access to these weapons. Even in the United States, civilians cannot buy high explosives or the sophisticated equipment to detonate them. Only the United States military and the Israelis have access to such ordnance.

Who might make up the 'B-Teams?' On the American side we have many secret and semi-secret units in the field not under the Army's chain of command. These are the CIA operatives, the semi-military "task forces, the Green Berets, the Seals, the Israeli Mossad and the 'security contractors' we employ. These last are pure mercenaries who will kill for money; when there is enough of it. In the British forces there is another organization known as the "Special Army Section" or SAS that performs similar roles.

Here is a recent report from Al Jazeera, the Arab television and news agency, describing how certain "suicide" bombings are being "arranged."

One morning an Iraqi farmer was driving with his young son in his pickup truck loaded with melons that he hoped to sell in the Baghdad marketplace. While driving down the road he was stopped by what appeared to be two armed American soldiers. They wanted to search him and his truck. One soldier searched him at the front of the truck while the other soldier searched through the melons in the back. After a few minutes they told him he could go on.

However, when he got back on the road his son said that the soldier at the back of the truck had placed something among the melons. It was about the size of a melon but it was a grey metal object. The farmer stopped and found it. It looked ominous to him and he at first did not know what to do with it. If he touched it, it might explode. But finally, because the truck to him was his only way of making a living, he decided to take a chance so he took the ball-like object in his arms and left it out in a nearby field.

When he came back from the market he went to look for it but found that it must have been a bomb because it had exploded. Happily, he said, it had only killed one goat.

This account sounds believable. Certainly dozens and hundreds of Iraqis are not bent on becoming suicide bombers. Perhaps their cars have been searched and also had bombs placed in them. Then, the newspapers next day report them as being suicide bombers who had exploded their cars in crowds of their own people. Could this scenario be the ghastly truth?

SAS

In a series of events that began on September 19th 2005. Two men dressed as Iraqis were driving in their car in the British occupied city of Basra. When Iraqi police attempted to stop them to ask what they were doing. The men in the car refused to comply and began shooting at the police. After an exchange of gunfire the two men were finally arrested and it was found that they were in reality two British SAS soldiers dressed in Iraqi disguise (turbans false beards and the lot) and that their car was filled with explosives and heavy weapons. With that the police took them and locked them up in the Basra city jail.

When higher British authorities heard of this, everything must have hit the fan; with the news going right on up to the Prime Minister, Tony Blair, in London. Within hours the *word* came back down that the two SAS men had to be recovered or exterminated before they talked and exposed their mission.

There was little negotiation. When the Iraqis would not immediately release the men the British Army mounted a full scale military assault upon the jail; knocking down the building with tanks. Afterward the SAS men were found to be unharmed and were returned to their unit. This story compliments the one above in showing that the British as well as the Americans have their own B-Team that is creating many of the so-called atrocities in Iraq that are then blamed upon the 'Insurgency.'

There may in fact be no 'Insurgency' or Iraqi "civil war" but only a fake war; our A-Teams vs. our own B-Teams; used to advance the purposes of our diabolical leaders.

*   *   *   *

SADMs

While every civilized person believes that atomic warfare must be avoided at all costs, the truth is that mini-nuclear bombs are already being employed by our B-Teams in the staging of fake "terrorist" bombings.

It has come to light that what our atom-bomb scientists have been doing over the last 60 years is not only designing large explosive devices but also they have been working to reduce the size of atomic bombs so they may be used as limited and controllable devices.

Exploding a hydrogen bomb requires a two stage process. In small munitions such as an artillery shell, the first stage is a small plutonium 239, device or trigger mechanism that initially fires and creates the necessary explosion and blast of radiation required to ignite the hydrogen material. A hydrogen bomb's explosion, of course, has a devastating effect.

But, in an effort to build smaller and more manageable demolition devices, it was decided to use only the first stage trigger device as, what are now called Special Atomic Demolition Munitions or (SADMs).

It is believed by some investigators that these SADM bombs have already been used in the Bali nightclub bombing that was directed primarily against Australian vacationers and in the bombing of the Australian Embassy in Jakarta, Indonesia.

This conclusion has been reached because an atomic explosion leaves a signature that a chemical explosion does not. The difference between these two is that in the case of a conventional chemical bomb, using TNT or C-4 material, the explosion is progressive and takes place over several milliseconds of time. If the bomb is on the surface, it explodes progressively in a whoosh, and its force goes outward, horizontally, and upward, vertically, but does not make a crater in the earth.

In contrast, a plutonium-239 SADM ignites, totally, in a millionth of a second. Anyone and anything within its range is not blown away; rather, it is atomized. It ceases to exist. It disappears. At the time of ignition, there is seen a flash of blue light caused by the Alpha radiation atomizing the atmosphere but there is no time for the effect to progress sequentially. Instantly, it vaporizes everything within its range including the earth below it. Therefore, it leaves a crater. That is its signature.

These, otherwise, inexplicable craters have been observed at the sites of both the Bali bombing and at the site of Australian Embassy bombing in Jakarta. It appears that these SADMs are also being used in fake "terrorist" attacks against helpless civilians in Baghdad.

Who can be supplying these SADMs? There are only two possible sources. One is the United States and the other is the Nation of Israel. Theirs is called the

"Dimona" bomb because it is being manufactured in the town of Dimona in the Negev desert of Israel.

*   *   *   *

News Report: Baghdad, October 31, 2006. Associated Press

An explosion occurred at a popular gathering point in the Shiite assembly area for day-laborers looking for a day's work. The blast spewed shards of metal, and blew up three nearby cars, <u>and left a huge crater in the pavement!</u>

"Shiite leaders accused the Americans of complicity in the market blast, saying that because they were in charge of searching all vehicles going in and out of the area, they must have, allowed in, the bomb that was detonated at the market. "They let it in and slaughtered the people."

This news report confirms that PAGAN is without mercy. It is beginning to wage an atomic war against the world's people that may end up destroying civilization!

*   *   *   *

THINGS TO COME

After the Federal Government's obvious staging of the 9/11 atrocities and the passage of the Patriot Act it is apparent that the United States has been infiltrated and taken over, totally, by the Jesuit's cabal: PAGAN. Politicians from the President on down along with the nation's so-called moral and intellectual leaders have, all together, declared that they are internationalists—citizens of the World and not Americans. Therefore the problem being faced by the remaining members of our former Republic is not how to fend off these infiltrators;—that's all over. The infiltrators are inside. The problem now is how to overthrow them and establish a new of government upon the ashes of the old.

The United States is now a National Socialist Nation. The passage of the two Patriot Acts marks the same transition here as did passage of the 'Enabling Act' by the German Reichstag in 1933. The Patriot Acts I and II have, by law, set up a National Socialist Dictatorship. The 'Homeland Security Administration' is the enforcement agency; it is the Jesuit "Gestapo."

WHAT HAPPENS NEXT IN AMERICA?

The Jesuits and their Jewish straw-men, the Neo-Cons, have established a totalitarian government over us but they know, all too well, that they only rule from the top of the pyramid; they do not control the hearts and souls of the people. They may rule for a time but their problem will be to make the lower ranks "believe" and "obey." This, they will attempt to do by binding the populace with the chains of fear; by isolating them one from another; by prohibiting free thought and by forbidding the exchange of information with censorship. As in Nazi Germany and in Soviet Russia, individuals will be given an ID cards and need permission to travel or to move from one city to another or to change jobs. Then, will come the "purges" and the concentration camps.

These are sobering thoughts but when this plan is recognized, the American people will emit a growl that will mark the rising of the people; as in the rising of the Protestant Reformation.

AND, AFTER THAT?

What occurs next is that Christians in small groups must mobilize themselves in educational action groups at the lowest level. Politically they will work from the bottom up; capturing school boards city and county governments until they can control the states. When blocked, they will infiltrate the camp of their enemies and convert them one by one. The ordinary government lackey is not essentially evil even though his government might be. He too loves truth and justice. Convert him and move on up the chain of command.

Do not imagine this will be a bloodless operation. The tyrants will do their best to root out religious opposition with every weapon they command. It will be a time for martyrdoms; but each example will win over many more until finally the PAGANS at the top will find they hold only nominal rule over a nation of militant Christians; who will no longer believe or obey.

The process may take generations but the outcome is certain. During this time governments will rot from within. Factions will develop. Assassination will become the method of choosing new leaders. Taxes will not be collected because the people will be unable to pay. Finally, the armies will seize control when, and as, the political leaders become impotent. The last days of the First Roman Empire will repeat themselves in modern dress.

*   *   *   *

THE CHRISTIAN RESPONSE

Christians must stand fast! Respond to the criminal deceptions of PAGAN by proclaiming the truth of the gospel of Jesus Christ. Do this to all who will listen and to those who, at first, will not listen.

In days to come, the true path for Christians will be to remember that they live in the Kingdom of God. Their role is to practice Christian Separation and to lead by example; living by God's Commandments and converting others. Never should Christians willingly support or participate in the wicked schemes of PAGAN. Christians should remember that the religious liberties they enjoy today were bought for them with the blood of the martyrs. We must stop PAGAN now or we, otherwise, will pay that price again.

Christians must rise up and proclaim the truth of the Gospel and provide others with a true understanding of history. The Christians of America are not surrounded by enemies; they are surrounded by millions of people who are confused but who are seeking the answers we possess. Those answers can be given to them by providing them with the true Bible and the Gospel of Christ. All these; Catholics Agnostics and whomever are fundamentally good people. They do not represent a problem; they represent an opportunity.

What we must do is to convert them; Catholics most especially. Many are ready for conversion. They are well aware that their man-made religion is flawed. They need only to be shown how and why they should renounce their belief in the snares of the Papacy; their belief in idols, in dead persons (the so-called saints), in relics and in their flawed priests and nuns. Then they will come gratefully; submitting themselves to Christ as their true high-priest whom they can approach and pray to directly. Catholics are already half way. It should be our duty and our joy to witness to them and to bring them into the Kingdom of God. This is the certain and the only way we can undermine and defeat Pagan while at the same time reestablishing our country as a Christian nation under God.

Here we must stand because like Martin Luther; we cannot do otherwise.

BIBLIOGRAPHY

Most books, out-of-print, are still available. Books printed since 1900 can be found through used book dealers. Search the American Book Exchange: It can be accessed on their web site: www.abe.com/http://www.abe.com/

*   *   *   *

Anonymous. *Report from Iron Mountain on the Possibility and Desirability of Peace.* New York N.Y.: Dell Publishing Company, 1967.

Carposi, George. *Clinton Confidential; Bill Clinton's Rise to Power.* New York, N.Y.: Emery Dalton Books, 1995.

Castillo, Bernal Diaz. *The Discovery and Conquest of Mexico,* New York, N.Y.: Farrar Strauss & Cudahay, 1956.

Chambers, Whitaker. *Witness,* New York, N.Y.: H. Wolff & Company, 1952.

Cornwell, John. *Breaking Faith*; New York, N.Y.: Viking Penguin Books, 2001.

Cornwell, John. *Hitler's Pope; the Secret History of Pius XII,* London/New York, N.Y.: Viking-Penguin Books, 1999.

Dall, Curtis B. *F. D. R. My Exploited Father-In-Law,* Washington, D. C.: Action Associates, 1970.

DeSoto, Hernando. *The Mystery of Capital,* New York, N.Y.: Basic Books, 2000.

Encyclopedia Britannica Eleventh Edition, Cambridge University Press, 1911.

Gibbon, Edward. *The Decline and Fall of The Roman Empire.* New York, Modern Library, 1985.

Grant, Ulysses S. *Personal Memoirs,* New York, N.Y.: Charles L. Webster & Company, 1894.

Herodotus. *The Histories of Herodotus,* Translated by George Rawlinson. New York, N.Y.: Tudor Press, 1928.

Hislop, Reverend Alexander. *The Two Babylons; Or the Papal Worship,* New York, N.Y.: Loizeaux Brothers Inc, 1916.

Hitler, Adolf. *My Battle,* Boston, Mass.: Houghton Mifflin Company, 1933.

Holy Bible; King James Translation, Cleveland, Ohio: World Publishing Company, 1950.

Horowitz, David. *Radical Son, A Journey Through Our Times*. New York, N.Y.: The Free Press, 1997.

Kelly, Virginia. *Leading With My Heart*, New York, N. Y. : Simon & Schuster, 1994.

Kessler, Ronald. *Sins of the Father; The Life of Joseph P. Kennedy*. New York: Warner Books, 1994.

Kiyonaga, Bina Cady. *My Spy; Memoirs of a CIA Wife*, New York, N.Y.: Avon Books, 2000.

Knuth, E. C. *The Empire of the City; World Superstate*. Milwaukee, Wisconsin: The Author, 1946.

Koestler, Arthur. *The Thirteenth Tribe, A History of the Kahzar Empire*. New York, N.Y.: Random House, 1976

Laake, Deborah. *Secret Ceremonies, A Mormon Woman's Intimate Diary of Marriage and Beyond*, New York, N.Y.: Island Bantam, Doubleday & Dell Publishing Group, 1993.

LePlongeon, Augustus. *The Mayas' and the Quiches' Sacred Mysteries*, New York, N.Y.: Robert Macoy, 1886.

Lundwall, N. B. *The Vision or the Degrees of Glory*, Salt Lake City, Utah: Author, Compiler and Publisher, 1944.

Manvell, Roger. *SS and Gestapo, Rule By Terror*, New York: Ballentine Books, 1969.

Martin, Malachai. *The Jesuits*, New York: Touchstone Books/Simon & Schuster, 1987.

McCabe, James D. *Cross and Crown or The Sufferings and Triumphs of The Heroic Men and Women Who Were Persecuted For The Religion Of Jesus Christ*. Philadelphia, Pennsylvania: Jones Brothers & Company, 1873.

McCullough, David. *Truman*. New York, N.Y.: Simon and Schuster, 1992.

McTaggart, Lynne. *The Field*, New York: Harper Perennial, 2003.

Menjou, Adolphe. *It Took Nine Tailors*. New York: Whittlesey House, Mc Graw Hill, 1948.

Kiyonaga, Bina Cady. *My Spy*. New York: Avon Books, 2000.

Remusat, Madam de *Memoirs of Madam de Remusat; Life in Napoleon's Court*, New York, N.Y.. Appleton & Company, 1879.

Seymour, Charles. *The Intimate Papers of Colonel House*, New York, N.Y.: Houghton Mifflin Company, 4 vol. 1926.

Spengler, Oswald, *The Decline of the West*, 2 vols. New York N.Y.: Alfred Knopf Inc., 1928.

Steers, Edward Jr. *Blood On The Moon;* Kentucky: University Press of Kentucky, 2001.

Stephens, John L. *Incidents of Travel in Central America, Chiapas and Yucatan.* New York, N.Y.: Harper Brothers, 1841. Re-printed by Dover Publication. Paperback 2 vols., 1969.

Sumption, Jonathan. *The Albigensian Crusade*, London, England: Faber & Faber Ltd., 1988.

The Student's Milton. New York, N.Y.: Appleton-Century, 1930.

Ten Boom, Corrie. *The Hiding Place*, New York: Bantam Books, 1974.

Tosches, Nick. *Power on Earth;* New York, N.Y.: Arbor House, 1986.

Walsh, William Thomas. *Philip II, King of Spain*, New York, N.Y.: Sheed and Ward Inc., 1937.

Webster, Nesta H. *The French Revolution*, Hawthorn, California: The Christian Book Club, 1969.

Wills, Garry. *Papal Sin; Structures of Deceit*, New York, N.Y.: Doubleday Publishing, 2000.

978-0-595-44629-2
0-595-44629-9

www.ingramcontent.com/pod-product-compliance
Lightning Source LLC
Chambersburg PA
CBHW030309290526
45785CB00001B/278